Social Tourism: Global Challenges and Approaches

Social Tourism: Global Challenges and Approaches

Edited by

Joana Lima

Social Sciences School, University of Évora and CIDEHUS Research Unit, Évora, Portugal

Celeste Eusébio

Department of Economics, Management, Industrial Engineering and Tourism, University of Aveiro and GOVCOPP Research Unit, Aveiro, Portugal

CABI

CABI is a trading name of CAB International

CABI
Nosworthy Way
Wallingford
Oxfordshire OX10 8DE
UK

Tel: +44 (0)1491 832111
Fax: +44 (0)1491 833508
E-mail: info@cabi.org
Website: www.cabi.org

CABI
WeWork
One Lincoln St
24th Floor
Boston, MA 02111
USA

T: +1 (617)682-9015
E-mail: cabi-nao@cabi.org

A catalogue record for this book is available from the British Library, London, UK.

References to Internet websites (URLs) were accurate at the time of writing.

ISBN-13: 9781789241211 (hardback)
9781789241228 (ePDF)
9781789241235 (ePub)

DOI: 10.1079/9781789241211.0000

Commissioning Editor: Claire Parfitt
Editorial Assistant: Emma McCann
Production Editor: James Bishop

Typeset by Exeter Premedia Services Pvt Ltd, Chennai, India
Printed and bound in the UK by CPI Group (UK) Ltd, Croydon, CR0 4YY

Contents

List of Contributors

Alén González, Elisa is a professor of marketing specializing in tourism marketing at the Faculty of Business Sciences and Tourism, University of Vigo. Elisa's research work lies in the area of tourist behaviour, ranging from motivation to behavioural intentions, with a focus on improving satisfaction. In recent years, she has focused on different types of tourism, mainly thermal and senior tourism. **alen@uvigo.es**

Alvelos, Helena graduated in electronics and telecommunications engineering, received an MSc degree in management (MBA) and a PhD in engineering sciences. She is currently an assistant professor in the Department of Economics, Management, Industrial Engineering and Tourism at the University of Aveiro, and teaches subjects in the fields of applied statistics and quality. She has published in several journals and conferences, and currently she is a researcher at the Centre for Research and Development in Mathematics and Applications (CIDMA). Her current research interests include data analysis, quality management and statistical quality control. **helena.alvelos@ua.pt**

Carneiro, Maria João is assistant professor of tourism at the University of Aveiro (UA) (in Portugal) and a researcher at the Research Unit on Governance, Competitiveness and Public Policies (GOVCOPP) at the same university. She holds a degree in tourism management and planning from the UA, an MBA from the New University of Lisbon and a PhD in tourism from the UA. She is vice-director of the degree in tourism management and planning at the UA. She has published papers in several journals and conference proceedings. Her research interests include consumer behaviour in tourism, accessible tourism, cultural tourism and tourism impacts. **mjcarneiro@ua.pt**

Carr, Neil is a professor in the Department of Tourism at the University of Otago and the managing editor of *Annals of Leisure Research*. His research focuses on understanding behaviour, welfare and rights within tourism and leisure experiences, with a particular emphasis on animals, children and families, and sex. **neil.carr@otago.ac.nz**

Diekmann, Anya is professor in tourism at the Université Libre de Bruxelles (Belgium) and co-ordinator of the Alliance for Research of ISTO (International Organisation of Social Tourism). Her research integrates aspects of social tourism in Belgium and Europe with a particular focus on social tourism systems and the benefits of tourism on well-being and social inclusion. **Anya.Diekmann@ulb.be**

Domínguez Vila, Trinidad is an associate professor of business management and marketing in the Faculty of Business Sciences and Tourism, University of Vigo. Her mainly research interests are focused on tourist behaviour and motivation of people with disabilities. Nowadays, her research works are focused on seniors and smart tourism too. **trinidad@uvigo.es**

Estrada-González, Adriana E. is a leisure/tourism professor at Universidad de Monterrey (Mexico). She received a BA in leisure management studies from UMNE (Monterrey, Mexico). She continued her studies at WICE (Leeuwarden, The Netherlands), holding an MA in international leisure studies, and she studied a PhD in leisure and human development at Universidad de Deusto (Bilbao, Spain). Adriana was a member of the World Leisure Organization (WLO) Board of Directors (1997–2003) and was in charge of the Local Secretariat of the 3rd WLO Congress (Jaipur, India); she also organized a WLO International Seminar 'Leisure Education and Youth at Risk' (Monterrey, Mexico). She was awarded the WLO Hillel Ruskin Memorial Scholar Lecture (2014). **adriana.estrada@udem.edu**

Eusébio, Celeste is associate professor of tourism and a researcher at the GOVCOPP (Research Unit on Governance, Competitiveness and Public Policies) at the University of Aveiro (Portugal). She holds a degree in tourism management and planning from the University of Aveiro, a Master's in economics from the University of Coimbra and a doctoral degree in tourism, also from the University of Aveiro. Her research interests include social tourism, accessible tourism, tourism economics, tourism impacts, tourism forecasts and consumer behaviour in tourism. She is at present the coordinator of the degree in tourism management and planning at the University of Aveiro. **celeste.eusebio@ua.pt**

Figueiredo, Elisabete is a sociologist and an associate professor, with habilitation, of the Department of Social, Political and Territorial Sciences and researcher at the GOVCOPP (Research Unit on Governance, Competitiveness and Public Policies) of the University of Aveiro. Her research interests include contrasting social representations and imaginaries on rural areas and on rurality; rural development policies and strategies; rural reconfiguration and restructuring processes; rural tourism; and social and accessible tourism. **elisa@ua.pt**

Gabruč, Janja is assistant professor at the Faculty of Tourism Studies – Turistica, University of Primorska in courses covering employees and consumers in tourism. She graduated from the College of Social Work in Ljubljana and completed her Master's and doctoral studies at the Faculty of Organizational Sciences of the University of Maribor. She has participated in several national projects. Her main research interest is social sustainability in tourism (social tourism). **Janja.Gabruc@fts.upr.si**

Kakoudakis, Konstantinos I. is a lecturer in hospitality and tourism management at the University of Central Lancashire, Cyprus Campus. His research interests revolve around social tourism, sustainable development and working conditions in the hospitality industry. He was previously researcher/consultant in tourism development projects co-funded by the European Union, while he has also several years of work experience in the tourism industry. He holds a PhD from the University of Nottingham, UK. **kkakoudakis@uclan.ac.uk**

Kastenholz, Elisabeth is associate professor in the Department of Economics, Management, Industrial Engineering and Tourism, University of Aveiro, where she teaches tourism- and marketing-related subjects and coordinates a PhD program in Marketing and Strategy. She holds a PhD in tourism studies, an MBA, a degree in tourism management and planning, and a bachelor's in public administration – specialty foreign affairs. She is a member of the Research Unit GOVCOPP, of the Deutsche Gesellschaft für Tourismuswissenschaft and of the Portuguese Society of Rural Studies. Her research interests focus on consumer behaviour in tourism, the tourist/destination experience, sustainable destination marketing, and development, rural, wine and accessible tourism. **elisabethk@ua.pt**

Komppula, Raija is professor of marketing, especially tourism business, at the University of Eastern Finland, Joensuu Campus, Business School. Her research interests include research on tourist experience, new service development in tourism, destination branding, small business research and entrepreneurship, cooperation and networks, particularly in the context of rural and nature-based tourism. The field of well-being tourism, with a focus on nature-based tourism and social tourism are current areas of interest of her research group. Professor Komppula has several positions of trust in the tourism industry in Finland, and she is a member of AIEST, ISTO and TRC. **raija.komppula@uef.fi**

Lima, Joana is an assistant professor at the University of Évora and a researcher at the CIDEHUS and GOVCOPP Research Units. She holds a PhD in tourism from the University of Aveiro. Her PhD thesis focused on the effects of family tourism, particularizing the case of the low-income families. Since 2008, she has been a research assistant in four research projects in Portugal, of which three were in the field of social tourism. She was the main research assistant in another research project aimed at studying the rural tourism experience and sustainable development. Her main research areas of interests are: social tourism, family tourism, impacts of tourism on tourists and the tourist experience. Her main publications are in the fields of rural tourism, accessible tourism and family tourism. **jisl@uevora.pt**

Losada Sánchez, Nieves is a researcher in the Centre for Transdisciplinary Development Studies (CETRAD), University of Trás-os-Montes e Alto Douro. She is a specialist in senior tourism. **nlosada@utad.pt**

Loureiro, Cátia holds a degree in tourism from the Polytechnic Institute of Gaya (2008) and a Master's in management and planning in tourism from the University of Aveiro (2014). **wypa_16@hotmail.com**

McCabe, Scott is professor of marketing and tourism at the Nottingham University Business School and professor of tourism marketing/management at the University of Eastern Finland, Joensuu. Scott's research is primarily on tourist consumer behaviour and tourist decision-making. More recently, his work has engaged debates in pro-social and pro-environmental tourist consumption, as well as responsible and ethical consumer behaviour. Scott is interested in the qualities of tourist experiences, and how experiences are designed and marketed. **Scott.McCabe@nottingham.ac.uk**

Medarić, Zorana is an assistant at the Faculty of Tourism Studies, University of Primorska, in courses covering the social aspects of tourism and qualitative research in tourism. She graduated in sociology at the Faculty of Social Sciences of the University of Ljubljana and received her Master's degree in sociology – social politics from the same faculty. She is also a researcher at the Science and Research Centre of Koper. She has been involved in several national and international projects in the fields of tourism, sociology of everyday life, interculturality and migration. **Zorana.Medaric@fts.upr.si**

Michel, Xavier is an assistant professor in geography at the University of Caen-Normandy (France), where he is responsible for the Master's degree in educational travel and tourism (Voyages, séjours, mobilités scolaires et éducatifs). His research interests are focused on outdoor tourist experiences, tourist mobilities, tourism and well-being, slow tourism, youth, social and educational tourism (summer camps, school trips and student mobilities). He deals in particular with the access to tourist experiences and tourist relationships with the environment. He is a member of the board of several French social tourism organizers. **xavier.michel@unicaen.fr**

Minnaert, Lynn is the academic director and a clinical associate professor at the Jonathan M. Tisch Center for Hospitality and Tourism at New York University. Her research interests are social tourism for low-income groups, family tourism and the social legacies of mega-events. Her research has been funded by the Economic and Social Research Council, the European Union, the International Olympic Committee, regional tourist boards and Meeting Professionals International (MPI). Since 2015, she authors the annual US Family Survey in collaboration with the Family Travel Association. She is on the editorial board of Hotel and Tourism Management, and on the international advisory board for the *International Journal of Contemporary Tourism Research*. **l.minnaert@nyu.edu**

Papadoulaki, Katerina is administrative operations supervisor in the Department of Mathematics and Applied Mathematics, at the University of Crete, Greece. She has previously worked at the Hellenic Open University, School of Humanities. Her research interests include alternative forms of tourism, innovation in tourism and tourism education, and entrepreneurship with an emphasis on the historical development of travel and tourism enterprises. She holds a PhD in social and economic history from the Hellenic Open University, School of Humanities, in Greece. **kpapad@uoc.gr**

Pedrosa, Mariana is a controller and management assistant at the Implantology Institute in Lisbon (Portugal). She has a degree in economics from the University of Aveiro (2012), Portugal, and a Master's degree in tourism management and planning from the same University (2016). In her Master's thesis the main areas covered were social tourism, families with disabled children and social tourism programmes for these families. **msnpedrosa@gmail.com**

Rutherford, Gill is a senior lecturer in education and disability studies at the University of Otago, New Zealand. A teacher educator, her research and practice focus on inclusive education (including inclusive teacher education); disability studies; disabled students' experiences of school; and teacher aides' role in education. **gill.rutherford@otago.ac.nz**

Shiraani, Fathimath is a PhD candidate at the University of Otago, New Zealand. Her PhD focuses on exploring the tourism experiences of disabled children, as described and understood from the perspectives of children. **fathimath.shiraani@postgrad.otago.ac.nz**

Vento, Elli is a project researcher and a PhD student at the University of Eastern Finland, Joensuu Campus, Business School. Currently, she is exploring the effects of social tourism in the Finnish context. **elli.vento@uef.fi**

Acknowledgements

This work was funded by national funds through the Foundation for Science and Technology, under the project UIDB/00057/2020.

Introduction

Joana Lima[1]* and Celeste Eusébio[2]
[1]Social Sciences School, University of Évora and CIDEHUS Research Unit, Évora, Portugal; [2]Department of Economics, Management, Industrial Engineering and Tourism, University of Aveiro and GOVCOPP Research Unit, Aveiro, Portugal

Why a Book Reflecting on Social Tourism Around the World?

The importance of holidays and travelling for human well-being has been formally recognized since 1948, with the recognition by the United Nations of holidays as a basic human right for individuals. This importance has since been reiterated by several governments that have recognized the right to paid holidays over the years (Hall and Brown, 2006). As stated by Richards (1999), holidays away from home (tourism) mitigate two constraints of daily life – time and place – creating conditions for individuals to dedicate themselves to their personal and social fulfilment.

Despite the recognition of tourism as an important part of present-day life, today's society is also marked by a worsening of social inequalities, even within developed countries. Consequently, a significant part (not to say a majority) of the population still doesn't have access to tourism. There are several constraints to the practice of tourism, among which the low level of income stands out as one of the most important.

In order to combat this inequality of access to tourism, the development of tourism programmes aimed at disadvantaged social groups is a possible measure to provide individuals who do not have access to tourism the opportunity to participate in this activity. Existing studies, although it is recognized that there is no unique definition, consensually consider that social tourism is related to initiatives that aim to include groups in tourism activities who are otherwise excluded from it. This measure, despite already arousing the interest of various entities (such as the European Economic and Social Committee, several private social tourism organizations and some governments around the world), has not yet received extensive and effective recognition or investment (either regarding financing or scientific studies).

As mentioned before, this is a topic that has only received more attention from researchers in recent years: only from 2008 did the number of publications start to grow significantly (according to a search performed on the Scopus database with 'social tourism' as keyword and limited to studies with this expression in the title, abstract or keywords), even though the first published article, to our knowledge, dates from 1951.

Most of the authors in the field recognize the positive effects of social tourism for its participants and society as a whole (e.g. Smith and Hughes, 1999; Minnaert, 2006; Minnaert *et al.*, 2009; McCabe *et al.*, 2010). The results of most studies stress the importance of social tourism for its participants, in terms of well-being and quality of life improvement, and also emphasize

*Corresponding author: jisl@uevora.pt

DOI: 10.1079/9781789241211.0001

benefits for learning and development of social behaviour. These results seem to indicate that encouraging tourism for disadvantaged groups may facilitate social inclusion, encouraging social solidarity and collective cohesion and leading to a decrease in poverty and exclusion. Another common conclusion of most studies is the recognition of the need for public intervention in social tourism initiatives, and the challenges this fact represents, in a period with important financial restrictions for most governments, are commonly recognized (Lima and Moura, 2017).

Reflections on social tourism, both conceptual and empirical, still seem to be necessary as a starting point for designing efficient strategies to promote what is now considered a basic right of individuals and a dimension that characterizes modern society and quality of life – access to tourism. Existing research provides solid ground to develop further research, but there is indeed a window of opportunity for significantly improving knowledge on this topic.

In this sense, it is important to stress again that further research work is required, calling for more wide-ranging and transnational comparisons. Consequently, the opportunity to bring together conceptual chapters and others with practical examples from all around the world and analysing different targets inspired us to produce this book.

Structure of the Book

As editors of this book we are excited to present this compilation of rich, diverse chapters on social tourism. The following 13 chapters bring together a variety of perspectives on social tourism around the world, illustrate many approaches that can be taken and point out directions for the future of social tourism.

Part II is composed of case studies, with three chapters discussing the historical evolution of the concept and implementation of social tourism in different parts of the world. This discussion also results in the identification of challenges and opportunities for the development of social tourism in these contexts. In Greece, as discussed in Chapter 1, social tourism has been a part of social welfare policy for almost four decades. Its practice, however, has seen significant fluctuations over time owing to a diversity of factors, but programmes are still operative and are almost exclusively supported by the public sector. In Mexico, the development of social tourism is relatively recent as it was only during the 20th century that tourism became more formally organized and involved public, private and social participation, as pointed out in Chapter 2. In Finland, although social tourism has a long tradition, it seems to have been neglected in sociopolitical discussions, as stated in Chapter 3. This chapter introduces a case study concerning social tourism in Finland and focuses on challenges and opportunities for the development of social tourism in this country.

Since the 1990s, studies have been published on the effects of tourist practice for visitors, highlighting the importance of the practice of tourism for individuals' personal and social development (e.g. Smith and Hughes, 1999; Hall and Brown, 2006; Minnaert, 2006; Minnaert *et al.*, 2009; Alexander *et al.*, 2010). These studies identify a set of factors, of individuals and tourism practices, that can influence the way these effects are felt. Consequently, different personal, social and economic conditions, as well as different motivations and different tourist experiences for individuals, lead to different effects or different intensities of these effects (Smith and Hughes, 1999; Shaw and Coles, 2004; Alexander *et al.*, 2010; Dolnicar *et al.*, 2012). When considering studies on the effects of social tourism on participants, one conclusion became evident: the participants of social tourism programmes seem to feel the effects of engaging in tourism activity more intensely (Smith and Hughes, 1999; Minnaert, 2006; McCabe *et al.*, 2010).

Part III (seven chapters) of the book reflects on tourism experiences lived by different target groups of social tourism. An ageing population is a reality that is having a great impact on developed countries. With this in mind, Chapter 4 analyses the INATEL Foundation social tourism programmes offered to seniors and the benefits that the seniors obtained from their participation in these programmes. Regarding the same target, social tourism programmes for seniors in Slovenia and Spain are presented in Chapters 5 and 6, respectively. These programmes are analysed in terms of the characteristics and benefits and their potential for future development. Chapter

7 changes the focus of the analysis to a group that is one of the target groups of social tourism initiatives that faces most travel constraints – low-income families with children with disabilities. This chapter focuses on the Portuguese situation. In Chapter 8, there is a discussion of how to consider children with disabilities in tourism, by listening to their voices regarding their needs and motivations. Also focusing on children, Chapter 9 presents a reflection on the effects of social tourism for the well-being and social inclusion of economically disadvantaged children. Children are the target considered again in Chapter 10, but associated with a different reflection: this chapter discusses the financial accessibility of current children's camps in France.

The last part of the book is devoted to practical challenges and policy implications. This part starts with a chapter that aims to provide relevant guidelines to develop social tourism programmes for low-income families, based on the interest of a sample of Portuguese low-income families in participating in social tourism programmes (Chapter 11). Chapter 12 provides a reflection on the roles of the public, private and voluntary sector in social tourism provision in Flanders, with a view to designing new business models for solving social problems. This part of the book ends with Chapter 13, where a critical reflection concerning social tourism policies is undertaken.

The last part of the book presents the Conclusions. This chapter aims at identifying the main contribution of each chapter, summarizing some challenges and identifying approaches for the future of social tourism around the world.

Hopes and Expectations for the Book

As mentioned above, many people are still prevented from participating in tourism activities owing to various constraints (e.g. physical disabilities, poor economic conditions, reduced mobility and social isolation). To overcome these constraints to travel, it is essential to develop innovative social tourism initiatives worldwide. It thus becomes imperative to think critically about the challenges and the approaches concerning social tourism initiatives offered around the world in order to facilitate access for disadvantaged groups (seniors, low-income families, children and youth, and people with disabilities) to tourism activities.

This book seeks to bring together contributions from scholars from different parts of the world, in the wide-ranging field of social tourism, adding value to the existing knowledge. However, we recognize that this is only a sample of the initiatives being undertaken globally. For example, neither the situation in Brazil, a country that has a long tradition in social tourism, nor the emerging Asian situation are discussed in this book. Nevertheless, it presents a comprehensive, multidisciplinary, international and future-oriented perspective to understand social tourism in different countries and continents.

As the book presents theoretical reflections on this concept and also examples of practical implementation for different targets, we are hopeful that it will truly contribute to extending knowledge in this field and encourage academic discussion and knowledge-sharing regarding the relevance of thinking of tourism as a dimension of present-day life, in order to: (i) think critically about the challenges faced by the sector and the relevance of promoting social tourism initiatives for disadvantaged groups of society; (ii) promote research that bridges theory and practice, permitting the identification of guidelines for more effective social tourism initiatives; and (iii) encourage future research in the field.

References

Alexander, Z., Bakir, A. and Wickens, E. (2010) An investigation into the impact of vacation travel on the tourist. *International Journal of Tourism Research* 12(5), 574–590. DOI: 10.1002/jtr.777.

Dolnicar, S., Yanamandram, V. and Cliff, K. (2012) The contribution of vacations to quality of life. *Annals of Tourism Research* 39(1), 59–83. DOI: 10.1016/j.annals.2011.04.015.

Hall, D. and Brown, F. (2006) *Tourism and Welfare: Ethics, Responsibility and Sustained Well-being*. CAB International, Wallingford, UK.

Lima, J. and Moura, A. (2017) Reviewing social tourism for future research. *Revista Turismo & Desenvolvimento* 27/28(2), 27–31.

McCabe, S., Joldersma, T. and Li, C. (2010) Understanding the benefits of social tourism: Linking participation to subjective well-being and quality of life. *International Journal of Tourism Research* 12(6), 761–773. DOI: 10.1002/jtr.791.

Minnaert, L. (2006) Social tourism: a potential policy to reduce social exclusion? The effects of visitor-related social tourism for low income groups on personal and family development. PhD thesis. University of Westminster, London, UK.

Minnaert, L., Maitland, R. and Miller, G. (2009) Tourism and social policy – the value of social tourism. *Annals of Tourism Research* 36(2), 316–334.

Richards, G. (1999) Vacations and the quality of life: Patterns and structures. *Journal of Business Research* 44, 189–198.

Shaw, G. and Coles, T. (2004) Disability, holiday making and the tourism industry in the UK: A preliminary survey. *Tourism Management* 25(3), 397–403. DOI: 10.1016/S0261-5177(03)00139-0.

Smith, V. and Hughes, H. (1999) Disadvantaged families and the meaning of the holiday. *International Journal of Tourism Research* 1(3), 123–133. DOI: 10.1002/(SICI)1522-1970(199903/04)1:2<123::AID-JTR146>3.0.CO;2-R.

1 Social Tourism in Greece: A Brief History of Development from the Interwar Years to the Covid-19 Era

Konstantinos I. Kakoudakis[1]* and Katerina Papadoulaki[2]
[1]School of Business and Management, University of Central Lancashire, Cyprus Campus, Larnaca, Cyprus; [2]Department of Mathematics and Applied Mathematics, University of Crete, Heraklion, Greece

1.1 Introduction

Greece is among those European Union Member States, such as France, Belgium and Portugal, that show high levels of active state support with regard to social tourism (e.g. Walton, 2013). This is not surprising given that Greece has a long tradition in social tourism. Although it is difficult to identify the exact starting point of the development of social tourism in the country, available references show that this can be traced back to the interwar years. In the early 1930s, the newly established Greek National Tourism Organization (EOT) organized 'Sunday excursions' for the public in order to boost domestic tourism (Katsigiannis, 2017). Over time, Greece's political, social and economic environment has nurtured other motives too, such as the stimulation of national pride, the manipulation of the masses and the enhancement of all citizens' well-being and quality of life. Thus, in line with other countries, social tourism in Greece has evolved from both public and private initiatives, encompassing different motives (see La Placa and Corlyon, 2014). Its contemporary form, however, has been associated with the rise to power of the Panhellenic Socialist Movement (PASOK), which, after forming the first Greek socialist government in 1981, emphasized the provision of social welfare entitlements, especially to the lower middle class (see Pagoulatos, 2003), thus creating fruitful ground for the parallel development of social tourism in Greece.

Despite this long history though, and unlike social tourism systems in other countries, such as Belgium, France, Portugal, Spain and the UK (see Minnaert *et al.*, 2012; Eusébio *et al.*, 2016), very little has been written about social tourism in Greece within the international tourism literature. In fact, available information about Greek social tourism is limited to a consultancy report, 'Calypso Study on Social Tourism in Greece' (see Détente Consultants, 2010), and a recent paper by Kakoudakis and McCabe (2018). With regard to Greek bibliography, social tourism research also remains scarce, and any available studies have not intersected international literature owing to language and translation barriers, an issue that has been identified by tourism scholars in relation to non-English studies (e.g. Diekmann and McCabe, 2011). Moreover, there is a lack of available and coherent data on Greek social tourism policies and programmes. As a

*Corresponding author: kkakoudakis@uclan.ac.uk

DOI: 10.1079/9781789241211.0002

result, these barriers do not facilitate a discussion on the practice of social tourism in Greece over time. As Diekmann and McCabe (2011, p. 418) point out, these limitations 'represent a gap in the transfer of knowledge between countries and between organisations and academia, which has reduced the level of overall impact of social tourism research on policy'.

While acknowledging the above limitations, the long history of social tourism practice in Greece, and its inclusion in the public policy agenda over the past several decades, makes it a particularly interesting case to study. Therefore, the present chapter attempts to bridge this gap by illustrating the process of social tourism development in Greece, from the interwar years until the present day. The chapter first sets the discussion within the context of the country's turbulent political, social and economic background, throughout most of the past century, which has exercised significant influence on the development of Greek tourism in general, and social tourism specifically. It then identifies and presents two main phases of social tourism development, highlighting important initiatives and key players that contributed to the incremental evolution of social tourism programmes in Greece, and also events that impeded their implementation and smooth running. Specific emphasis has been given to the past four decades, since this time period has largely shaped the contemporary form of Greek social tourism programmes. Therefore, the chapter explicates the close linkages between the establishment of the modern Greek welfare state in the early 1980s, and the development of social tourism as we know it today. The chapter concludes with a brief discussion on the developmental process of contemporary Greek social tourism over time, and the important socioeconomic implications of its current practice in the aftermath of the Greek financial crisis, and in the midst of the refugee crisis in Europe, and the Covid-19 pandemic.

1.2 The Evolution of Social Tourism in Greece

Looking at the evolution of social tourism in Greece, we can generally identify two main phases of development, according to the degree of organization, application and continuity of social tourism initiatives. The first covers the period from the interwar years until the early 1980s. It is characterized by important and, at the time, novel initiatives, which created the ground upon which contemporary social tourism was founded. Despite their significance, however, these early initiatives lacked systematic organization and continuity. The second phase began in the early 1980s and continues today. It coincides with the establishment of the modern Greek welfare state and the recognition of tourism as a social right, leading to the design of contemporary social tourism programmes, and the expansion of social tourism practice in the country. Despite their differences, however, what both phases have in common are frequent political, economic and social instabilities, which have resulted in recurrent changes in the identity and structure of tourism public administration bodies (Table 1.1), affecting the implementation of Greek tourism policy in general, and social tourism policy and practice, in particular. Arguably, the degree of significance of these changes differs greatly, resulting in respective differences in their implications for social tourism practice. What is certain, though, is that this instability has resulted in severe fluctuations in the organization and direction of Greek tourism, including social tourism, over time.

1.2.1 Political, social and economic background

From a chronological point of view, the first phase of social tourism development in Greece took place during particularly turbulent years, characterized by major historical events (e.g. the Asia Minor catastrophe and its socioeconomic consequences, the Greek–Italian War and the military junta in 1967–1974), and constant political change. During the interwar years, Greece suffered from many political, social and economic problems, such as unstable governance, a large volume of Asia Minor refugees and financial crises (e.g. Forster, 1958). It is because of the latter that tourism attracted government attention and was viewed as a tool for development. Eleftherios Venizelos played a significant role in

Table 1.1. Recurrent changes in Greek tourism administration bodies (1929 – present).

Date	Events
1929	Establishment of the Greek National Tourism Organization (EOT), under the supervision of the Ministry of Economy
1935	Establishment of the Supreme Council of Tourism Abolishment of the Supreme Council of Tourism during the same year
1936	Establishment of the Deputy Ministry of Press and Tourism, which takes over the supervision of EOT Abolishment of EOT and transfer of its responsibilities to the Deputy Minister of Press and Tourism
1941	Abolishment of the Deputy Ministry of Press and Tourism
1944	Reestablishment of the Deputy Ministry of Press and Tourism
1945	Abolishment of the Deputy Ministry of Press and Tourism. Supervision of tourism-related services by the president of the government Establishment of the General Secretariat of Tourism
1946	Transfer of the responsibilities on tourism-related matters from the president of the government to the minister of coordination. Supervision of the General Secretariat of Tourism by the minister of coordination
1950	Abolishment of the General Secretariat of Tourism Reestablishment of EOT, under the supervision of the minister of the government's presidency
1951	Transfer of the responsibilities on tourism-related matters from the president of the government to the minister of national economy. Transfer of the responsibilities of the General Secretariat of Tourism to the secretary general of EOT
1956	Creation of the position of a specialist tourism consultant under the supervision of the president of the government
1965	Establishment of the Supreme Advisory Council of Tourism, under the supervision of the minister of the government's presidency
1978	Restructuring of EOT's services
1984	Transfer of the tourism-related responsibilities of the minister of the government's presidency to the secretary general of EOT
1989	Establishment of the Ministry of Tourism, which supervises EOT
1991	Abolishment of the Ministry of Tourism, and transfer of its responsibilities to the Ministry of National Economy
1993	Reestablishment of the Ministry of Tourism
1996	Merger of the Ministry of Tourism with the Ministry of Development
1997	Establishment of the National Council of Tourism
2001	Restructuring of EOT. Establishment of the Department of New Forms of Tourism Development (including social tourism)
2004	Establishment of the Ministry of Tourism Development as an independent ministry
2009	Merger of the Ministry of Tourism Development with the Ministry of Culture Establishment of the new Ministry of Culture and Tourism
2010	Supervision of EOT by the Ministry of Culture and Tourism
2012	Reestablishment of the Ministry of Tourism as an independent ministry

Continued

Table 1.1. Continued

Date	Events
2015	Merger of the Ministry of Tourism with the Ministries of Economy and Development, respectively Establishment of the new Ministry of Economy, Development and Tourism
2016	Reestablishment of the Ministry of Tourism as an independent ministry
2018	Main responsibility of EOT becomes the promotion of Greek tourism

Adapted from EOT (2012), Greek Government Gazette (1996, 2001, 2009, 2015, 2016, 2018), Raptarchis (2007).

the development of Greek tourism through several initiatives, such as the establishment of EOT (Papadoulaki, 2011). In addition, the first social tourism concepts in Greece emerged during his second administration (1928–1932), when EOT was considering ways to boost domestic tourism (Katsigiannis, 2017). The emphasis on tourism development continued during the dictatorship of Ioannis Metaxas, who took over the country's administration in 1936, as a result of national and international conjunctures. The first social tourism-related law (Law No. 2436/1940) was passed under his regime (Greek Government Gazette, 1940).

The liberation of Greece from the German occupation was followed by the civil war of 1944–1945. The next two decades after the civil war comprised a reconstruction period for the country, which was signalled by a shift from an agricultural economy to one more industrialized (see Miller, 2009). As Forster (1958, p. 229) explains, this shift resulted in 'both unemployment and underemployment, contributing to the political instability of Greece ... The political instability arising from these economic conditions was much increased by the Greek multi-party system which gives rise to short-lived governments.' At the time, the main objective of the Greek prime minister, Konstantinos Karamanlis, was a 'rapid expansion' of industry (e.g. steel and mining) and the tourism sector (see Miller, 2009). In the 1960s, Greece started showing some signs of economic prosperity, but new political turbulences in the mid-1960s resulted in a military coup and a 7-year dictatorship (1967–1974). After the fall of the military junta in 1974, the Greek Republic was restored and Karamanlis, leader of the liberal-conservative party Nea Democratia, took office until 1980 (Pagoulatos, 2003). One year later, the country joined the European Community.

A critical chronological point of the second phase is the early 1980s, when the modern welfare state was established in the country. The development of contemporary social tourism in Greece has evolved in parallel with the development of the Greek welfare state and has revolved around the institution of social rights. The term welfare state refers to a society that puts an emphasis on the well-being of its citizens through specific interventions, such as the provision of economic and social benefits (Goodin, 1988; Lowe, 1993). A fundamental element of the welfare state is the recognition on behalf of the state that citizens have specific social rights, the exercise of which is significant in the improvement of their quality of life (e.g. Cox, 1998). In line with this recognition, the state also recognizes its responsibility to enhance citizens' well-being and quality of life (Mishra, 1981) through various means, including the provision of social welfare. With regard to tourism, the Greek socialist government at the time recognized that the welfare state has not only the responsibility to provide less privileged citizens with the rights to education, healthcare and employment, but also with the right to holidays (Tourismos & Economia, 1986). Hence, publicly funded holidays in Greece have been realized as a result of this recognition (see Igoumenakis *et al.*, 1998), which is linked to the concept that going on holiday is an important indicator of social well-being (e.g. Haukeland, 1990; Kakoudakis *et al.*, 2017).

Although the concept of the welfare state dates back to post-war (1945–1975) times for most Western societies, in Greece it was synthesized much later, in the early 1980s, with the first coordinated initiatives for the implementation of organized social policies (see Oikonomou, 1996, pp. 1–2). It has been argued that Greece's contemporary welfare history coincides with the rise to power of PASOK, which formed the first Greek socialist government in 1981 (Tsakalotos, 1998). Indeed, social tourism is inherently linked with the social democratic movements (e.g. Diekmann and McCabe, 2011), and in turn with the concepts of the welfare state and social rights (e.g. Richards, 1998). The welfare state is a major political goal of social democracy (e.g. Klitgaard, 2007) and social rights a fundamental element of the welfare state (e.g. Cox, 1998). These strong linkages largely explain why social tourism practice has a long history in regions with traditionally more social democratic ideologies (e.g. continental Europe) (Richards, 1998). Within the specific context of Greece, this also explains why social tourism became a consistent part of the country's social welfare policy during PASOK's administration.

Despite the fact that the institution of the welfare state had been questioned at an international level during this period (Oikonomou, 1996), PASOK, led by Andreas Papandreou, largely focused its political agenda on social welfare. As Pagoulatos (2003, p. 95) explains, 'the rise of welfare entitlements in the first half of the 1980s served to compensate lower-middle-class strata for much of their real income losses due to the inflationary policies of the past years'. In so doing, the socialist government aimed to stabilize and strengthen its political grip by appealing to socioeconomically marginalized groups (Clogg, 1993). On the other hand, the rapid expansion of the Greek welfare state also benefited a large volume of citizens who were less in need, threatening the economic sustainability of social welfare provisions, including social tourism programmes (Igoumenakis *et al.*, 1998). Irrespective of any bad practices and paternalist political motives, it was during this period that living standards were improved for the majority of the Greek population for the first time (Pagoulatos, 2003), resulting in the democratization of tourism. In parallel, social tourism services were properly established and incorporated into Greece's welfare policy.

1.2.2 Phase one: From the interwar years to the early 1980s

In 1931, EOT's Board of Directors assessed the need to boost domestic tourism owing to the financial crisis that had hit international tourism (Katsigiannis, 2017, p. 272). At the time, EOT had put an emphasis on the development of spa towns, and the organization of Sunday excursions for the public, with the cooperation of national railways (Katsigiannis, 2017, p. 272). During the same period, 'Emporiki' bank offered its employees special leave and subsidized visits to spa towns (Dritsas, 2008), thus acknowledging that tourism participation is linked to workers' health, well-being and productivity (e.g. Minnaert *et al.*, 2012). Perhaps this vacation allowance for specific occupation groups was also an early recognition that those in paid employment have some sort of 'right' to a holiday (Minnaert *et al.*, 2006, p. 8). During Metaxas' regime, the Greek state, following the trends from other, more developed, European countries (e.g. Italy and Germany), offered workers significant travel discounts on rail and boat tickets and accommodation establishments through tourism coupons. These initiatives aimed to boost domestic tourism during a period of international military crisis (see Papadoulaki, 2011, pp. 97–98). In addition, Metaxas' approach to tourism had paternalist motives that aimed to stimulate national pride and manipulate the working class through state benefits, in the form of tourism and leisure (see Vlachos, 2013). This approach resembles 'a key historical component of 'social tourism' promotion by the state, in Europe and elsewhere' (Walton, 2013, p. 57).

References about social tourism in Greece in the late 1950s and 1960s are scarce. In 1959, a pilot social tourism programme was implemented on Lesvos Island (Nikolakakis, 2013). During the 1960s, social tourism practice in Greece ranged from programmes run by the Workers' Social Benefits Organization (Ergatiki Estia)[1] (a workers' fund supervised by the Ministry of Labour and Social Insurance) and campsites by the Ministry of Social Welfare and other organizations (e.g. banks, the former Social Insurance Organization [IKA] and the Hellenic Telecommunications Organization [OTE]), to initiatives organized by large enterprises (e.g. Piraiki – Patraiki and Olympic Airways) (Logothetis, 1967, p. 302). During the early years of Metapoliteusi (transition to democracy), and more specifically, in 1974, Ergatiki Estia also organized a 1-year pilot social tourism scheme with the cooperation of the Hellenic Chamber of Hotels, offering free holidays to 2000 workers

(Tourismos & Economia, 1974).[2] This trial programme covered the full costs of hotel accommodation and food for each participant worker and at least two members of his/her family for up to 10 days. The main objectives of this initiative were to boost domestic tourism, make holidays accessible to all social classes and establish social tourism in Greece, achieving continuity in the long term.

By the late 1970s, Greek tourism stakeholders and the state had acknowledged that social tourism could be seen both as an economic development strategy and a tool to exercise social policy (Tourismos & Economia, 1986). This renewed interest is confirmed by several significant initiatives. In 1976, for example, the 'Directorate H' Domestic – Social Tourism was established, under the supervision of the General Directorate of Tourism Development (Raptarchis, 2007). The Department II2 – Social Tourism of the Directorate, in cooperation with the 'Directorate A' Research and Development, had the responsibility of setting the necessary requirements for the development of social tourism. The Greek government at the time also considered it advisable to establish knowledge transfer partnerships with countries that had well-developed social tourism systems, such as France, in order to develop Greek social tourism upon solid foundations (Centre of Planning and Economic Research [KEPE], 1987). Indeed, in 1981, EOT signed a cooperation protocol with a French social tourism association; however, this ended soon after (KEPE, 1987).

1.2.3 Phase two: From the early 1980s to the present

The previous section shows that until 1980, the initiatives regarding social tourism appear to reflect single cases that lacked coherence and continuity, both with regard to their organization and implementation. Social tourism programmes in Greece started taking their current form in 1982, when EOT, Ergatiki Estia and the General Secretariat for Youth first launched social tourism programmes as we know them today (e.g. Igoumenakis *et al.*, 1998). EOT's first social tourism programme was also a pilot programme, which ran from 1982 until 1983 in cooperation with Ergatiki Estia and with the participation of EOT-owned hotels ('Xenia' Hotels) (Katsigiannis, 2017). The programme offered some valuable conclusions in relation to beneficiaries and providers. With regard to the former, it identified the specific social groups that should participate in the first social tourism programmes; with regard to the latter, it demonstrated the necessity of cooperation with other public and private organizations and trade unions (Tourism Issues, 1983). Since then, social tourism programmes in Greece have been largely organized and run by public authorities with the close cooperation of other public (e.g. Greek National Railways [OSE], the Common Collection Funds of Coaches [KTEL]) and private (hotels, campsites, and cruise lines) organizations, forming part of social policy (Igoumenakis *et al.*, 1998).

These partnerships have significantly contributed to the establishment of social tourism in Greece as a successful institution (Katsigiannis, 2017). This is confirmed by the available budget and large volume of beneficiaries. For example, during the first implementation period, until the mid-1980s, it is estimated that social tourism received 1.5 billion drachmas (about €4.4 million) of public funding, and benefited approximately 700,000 individuals and numerous businesses across the country (Tourismos & Economia, 1986). In addition to the main publicly funded social tourism programmes, some other important initiatives can be identified within the framework of wider social policy and trade union activities, such as vacation settlements for specific occupation groups (e.g. bank clerks, teachers and Greek army officers) and campsites for specific sociodemographic groups (e.g. the elderly, through the Open Care Centres for the Elderly [KAPI], and children with physical and mental disabilities, through the Patriotic Institution of Social Welfare and Awareness [PIKPA]) (see Sotiriadis, 1994).

From the early 1980s until the mid-2010s, EOT and Ergatiki Estia had been the main social tourism providers in Greece.[3] EOT also played a significant role in raising public awareness of social tourism and its socioeconomic benefits, and in developing more varied social tourism programmes, which aimed to expand the provision of tourism services to different sociodemographic groups. For example, in 1986, EOT's president chaired the International Bureau of Social Tourism Conference, held in Athens, which revolved around the economic and social implications of social tourism (Tourismos & Economia, 1986). Moreover, from the mid-1980s until the early 1990s, EOT launched social tourism subprogrammes, such as a winter youth tourism programme in popular ski centres; a therapeutic tourism programme in spa towns around Greece; and a 7-day holiday in Cyprus[4] (see Katsigiannis, 2017). Ergatiki Estia had a fundamental role, offering several social tourism programmes for almost three decades – from 1983 until 2012 (OEE, n.d).

On the other hand, and in spite of this significant progress during the 1980s, the practice of social tourism in Greece has not been smooth. During most of the 1990s, the country's economy was stagnant (for several reasons, e.g. budget deficit, low investment and high inflation), and under the supervision of the European Commission and the International Monetary Fund (IMF), which aimed to ensure reduction of the budget deficit (Close, 2002). These circumstances had significant negative effects on the Greek welfare state, and on the living standards of much of the population (e.g. restrained wages and social benefits, and increased taxation for prolonged periods) (Pagoulatos, 2003). Consequently, this climate of economic austerity had major implications for social tourism programmes. By the late 1980s, public funding was reduced, resulting in the shrinkage of EOT's programmes and, in turn, in significant reductions in the number of beneficiaries (Sotiriadis, 1994). Despite these adverse conditions, some important public initiatives can be still identified. In 1990, for example, EOT established the third General Directorate of Administration and Capitalization, and its Sub-Directorate of Domestic and Social Tourism ('Directorate H'), which supervised the implementation of social tourism programmes (Raptarchis, 2007).

The Greek economy began to recover in the mid-1990s. Indicative of this is that in 1995–2000, Greece became one of the best performers in the EU, reaching an annual average of 3.1% growth rate in GDP (Pagoulatos, 2003). In turn, this economic climate appeared to have positive implications for the practice of social tourism programmes in Greece. For example, in 1995, the provision of subsidized holidays for pensioners of the Greek Bar Fund (who did not receive pension from any other social fund) was initiated (Presidential Decree No. 225/1995) (Greek Government Gazette, 1995). The period of fast economic growth continued after Greece joined the euro zone in 2000–2008 (Matsaganis, 2011), and brought about positive changes in various sectors of the national economy, including tourism. Within this positive economic climate, the practice of social tourism thrived. In 2002, the Farmers' Welfare Account (Agrotiki Estia) was established within the Agricultural Insurance Organization (OGA) with the aim of organizing and implementing social tourism programmes for farmers, retired farmers and pensioners of the Non-Insured Elderly Account (Law No. 3050/2002) (Greek Government Gazette, 2002). In 2007–2008, EOT implemented its newly named 'Tourism for All' programme, including a specific subprogramme for seniors. During the first years of its implementation, the programme had a €15 million annual budget (Détente Consultants, 2010). Other important social tourism players during this period were the General Secretariat for Youth, unions and other insurance organizations (e.g. IKA, and the Mutual Health Fund of National Bank of Greece Personnel [TYPET]) (Détente Consultants, 2010).

The 2010s have been a period of deep economic recession for Greece, with significant consequences for society in general, and the Greek welfare state specifically (e.g. Matsaganis, 2011). As a result, social tourism practice has not remained unaffected. In 2012, one of the main and most successful providers of social tourism programmes in Greece, Ergatiki Estia, was shut down, affecting numerous citizens and hospitality SMEs (Kathimerini, 2012). In 2013, all the rights and responsibilities of Ergatiki Estia, including the provision of social tourism programmes, were transferred to the Manpower Employment Organization (OAED) (Law No. 4144/2013) (OAED, n.d). In 2014, EOT's 'Tourism for All' programme (including its sub-programme for seniors), the flagship social tourism programme in Greece, stopped running, and since then, to the best of our knowledge, it has been inoperative.[5] During the last period of its implementation in 2013–2014, the programme's budget had dropped to €5 million (EOT, 2013). It can be argued that these two events marked the end of an era for the two major providers of Greek social tourism for more than three decades. During the same year, the main responsibility for the design and implementation of social tourism programmes was transferred to the Department of New Forms of Tourism within the Ministry of Tourism (Greek Government Gazette, 2014).

Current programmes

Today, social tourism in Greece is largely organized and run by OAED and the Organization of Welfare Benefits and Social Solidarity (OPEKA, formerly OGA), while also encompassing smaller players, including unions, as well as other public and private organizations. Bank employees' associations, for instance, organize summer camps for their members' children (e.g. TYPET, 2018). With regard to the two major social tourism programmes run by OAED and OPEKA, these operate all year round, and mainly offer partially funded holidays, through coupons (subsidies), to different sociodemographic groups, such as employed and unemployed individuals, pensioners, people with disabilities (the carers of people with 67% disability or more are also eligible to participate) and their protected family members (OAED, 2018a; OPEKA, 2018). The eligibility criteria vary to some extent, depending on the provider; however, what all public programmes have in common is that their beneficiaries live on low incomes and have not participated in any other social tourism programme over the past year. With regard to accommodation suppliers, all types of hotels, rental rooms and campsites around Greece with a valid 'Special Operation Logo' are eligible to participate. Moreover, special emphasis has been given to the equal treatment of social tourists, through explicitly set responsibilities for hotel managers, in relation to the provision of good quality services and facilities (see OPEKA, 2018).

To a larger or lesser extent, the duration of holidays, the number of beneficiaries and the amount of subsidies differ among providers and programmes. For example, OAED's 2018–2019 programme had a €10 million budget, and offered up to five overnight stays to 140,000 beneficiaries, and up to ten overnight stays to those who choose to spend their holidays on the islands of Lesvos, Chios, Samos, Leros and Kos (OAED, 2018a). The aim of the latter initiative was to boost tourism activity in the North Aegean Islands, which have experienced a reduction in tourist demand owing to the refugee crisis in the region. This action resembles the practice of successive French governments, who have used the 'Chèque-Vacances' system to stimulate tourism activity in specific areas (see Davidson and Maitland, 1997). In addition, OAED runs a 'Camp for Kids' programme, with a €35 million budget, aiming to benefit 70,000 children (OAED, 2018b). The other major provider, OPEKA, offers social tourism holidays, through the Farmers' Welfare Account (Agrotiki Estia). Its 2018–2019 programmes included: 'Social Tourism' – up to 6 days for 55,000 beneficiaries, 'Therapeutic Tourism' – up to 6 days for 4000 pensioners, 'Camping Programme for Kids' – up to 16 days for 1000 children of beneficiaries and 'Four-Day Trips' for 12,500 beneficiaries (OPEKA, 2018). OPEKA also has a special offer (full accommodation subsidy, excluding breakfast) for those beneficiaries who choose to visit the North Aegean Islands and the Dodecanese. During the Covid-19 pandemic, and owing to its massive impact on international arrivals, the Greek Government has put a special emphasis on social tourism, as a means to boost domestic tourism (Greek Government, 2020). Right after the end of lockdown in the country, the government announced the implementation of a strengthened 'Tourism for All' programme for 2020–2021, through OAED. The budget of the programme has been significantly increased, reaching €37 million, and offering 370,000 vouchers (164% more compared with the previous year) (ETHNOS, 2020). In so doing, the Greek state has increased the amount of subsidy (while decreasing the amount of individual contribution), and the number of overnight stays from 5 to 6, has offered up to 10 overnight stays in Lesvos, Chios, Samos, Leros, Kos and Evros Regional Unit (with zero individual contribution), and has subsidized ferry tickets for the first time (OAED, 2020).

1.3 Conclusion

Social tourism is not a new concept in Greece. On the contrary, it has a long tradition, which can be traced back to the interwar years, when EOT was considering ways to boost domestic tourism during a period of international crisis. It has been mainly supported by the public sector, with the significant contribution of private organizations and labour unions, and it has seen growing interest, especially since the early 1980s when it was incorporated into the country's social welfare policy (Katsigiannis, 2017). However, this long history is rather unknown at an international level, as the case of Greek social tourism has not intersected tourism literature, mainly owing to the scarcity of relevant research and language barriers alike. This chapter has aimed to bridge this gap by presenting and discussing the development of social tourism in Greece over time, from its emergence during the interwar years until today. In so doing, it has become evident that Greek social tourism has been developed within a turbulent political, economic and social environment, largely characterized by instabilities and recurrent political changes. This environment has shaped the underlying motives for its development, and has exercised significant effects on its organization and practice (e.g. fluctuations in public funds, different priorities of successive governments in relation to public expenditure and, consequently, changes in the number of providers, programmes and beneficiaries, and the length of subsidized holidays). This confirms that social tourism is not a static but a dynamic concept (Minnaert *et al.*, 2012), and, therefore, any investment in social tourism is analogous to the political, economic and social circumstances of different regions at different periods of time.

It was identified that a plethora of initiatives, from the interwar period until the early years of Metapoliteusi, created the ground upon which contemporary social tourism is founded in Greece. However, it was in the early 1980s that Greek social tourism took its current form, becoming part of social welfare policy and consequently becoming more widely practised. This occurred in parallel with the rise to power of the first Greek socialist government, in 1981, and the establishment of the modern Greek welfare state. Indeed, it was during this period, when the connection between social tourism, social welfare and social rights became stronger in Greece, that a more systematic approach to social tourism development emerged. Although PASOK's administration had paternalist motives in relation to the implementation of social welfare policies, the wide practice of contemporary social tourism in Greece has evolved around the institution of social rights, and state recognition that less privileged citizens also have the right to holidays, among other important rights (Tourismos & Economia, 1986). In addition, the aim to boost domestic tourism has remained another major motive that still mobilizes the organization and application of social tourism programmes, emphasizing a more equal distribution of tourist flows throughout the year, and across regions. This has been especially pertinent in the aftermath of the Greek financial crisis, as well as in the midst of the current refugee crisis in Europe, and the Covid-19 pandemic.

Undoubtedly, Greek social tourism has been largely affected by these major phenomena, but, in parallel, it has also been addressing them through new initiatives. For example, the climate of economic austerity in Greece over the past few years has had significant implications for social tourism practice (e.g. abolishment of key providers and major programmes, such as Ergatiki Estia and EOT's 'Tourism for All' programme). In spite of these challenges, however, Greek social tourism has proved resilient, and continues to directly benefit thousands of disadvantaged individuals and accommodation businesses every year (e.g. OGA, 2017; OAED, 2018a). Today, it is mainly provided by two public organizations, OAED and OPEKA, while also encompassing smaller-scale initiatives, including programmes run by private organizations (e.g. banks). Specific emphasis has been placed on social tourism programmes for children due to the severe consequences of the financial crisis (e.g. high unemployment, acute poverty, risk of poverty and social exclusion) for a large proportion of Greek households (Hellenic Statistical Authority [ELSTAT], 2016). In the same manner, an emphasis has been recently put on the promotion of host communities that suffer from low tourism demand due to the refugee crisis in the region (e.g. the North Aegean Islands) as destinations for social tourists (OAED, 2018a). Most importantly, social tourism has been playing a major role during the Covid-19 pandemic. In this unprecedented global crisis that has paralysed international tourism, social tourism has been placed at the centre of Greece's tourism strategy, as one of the few remaining developmental tools with sustainable characteristics that have the potential to support domestic tourism. Arguably, these initiatives are of the utmost importance, as they address (at least partly) some of the major current socio-economic problems of Greece, including those related to the Covid-19 pandemic. Considering the complex and multidimensional economic, social and public health impacts of major crises (e.g. decline in tourism demand, increase of unemployment rates and increased risk of mental ill-health), forms of social welfare such as social tourism can offer some relief to vulnerable social groups and businesses, alike (Kakoudakis, 2020). Therefore, in this new and particularly turbulent environment that has been shaping over the past decade, and especially since the Covid-19 outbreak, the state has the responsibility to safeguard social tourism, and to take advantage of its full potential. For example, initiatives that will increase public awareness of the multiple benefits of social tourism, public–private partnerships and the inclusion of more providers (e.g. tourist and travel agencies) in future programmes, are deemed necessary towards this direction, and for the long-term sustainability of social tourism in Greece.

Notes

[1] Ergatiki Estia did not receive funding from the state budget, but its resources came exclusively from workers' and employers' contributions (OEE, 2012).

[2] The programme was organized during the Greek military junta, and it was probably launched sometime just before its fall, in July 1974. The article referred to was issued in July–August, 1974, and it includes an interview with Mr Paparrodopoulos, Minister of Labour during the first half of 1974.

[3] EOT had the responsibility for the beneficiaries working in the public sector, and Ergatiki Estia for those working in the private sector (see Sotiriadis, 1994).

[4] The programme had a budget of 55 million drachmas, and benefited thousands of Greeks (see Katsigiannis, 2017).

[5] To confirm this, we contacted the headquarters of EOT by phone.

Acknowledgements

We would like to thank Dr Vicky Story, professor of marketing and director of MBA programmes at Loughborough University, for taking the time to read our manuscript and give us her valuable feedback.

References

Clogg, R. (1993) *Greece 1981–1989: the Populist Decade*. Macmillan, London.

Close, D.H. (2002) *Greece Since 1945: Politics, Economy and Society*. Pearson Education, London.

Cox, R.H. (1998) The consequences of welfare reform: how conceptions of social rights are changing. *Journal of Social Policy* 27(1), 1–16. DOI: 10.1017/S0047279497005163.

Davidson, R. and Maitland, R. (1997) *Tourism Destinations*. Hodder and Stoughton, London.

Détente Consultants (2010) Calypso study on social tourism: Greece. European Commission. Available at: http://ec.europa.eu/DocsRoom/documents/6936?locale=en (accessed 23 November 2018).

Diekmann, A. and McCabe, S. (2011) Systems of social tourism in the European Union: a critical review. *Current Issues in Tourism* 14(5), 417–430. DOI: 10.1080/13683500.2011.568052.

Dritsas, M. (2008) *Emporiki Bank 1907–2007: Εναλλαγές ταυτότητας και μετασχηματισμοί [Emporiki Bank 1907–2007: Identity changes and transformations]*. Emporiki Bank of Greece, Athens.

ELSTAT (2016) Greece in figures: October–December 2016. Available at: www.statistics.gr/en/greece-in-figures (accessed 21 November 2018).

EOT (2012) Ιστορία του Οργανισμού [Organization's history]. Available at: www.gnto.gov.gr/el/history (accessed 1 December 2018).

EOT (2013) Πρόγραμμα «Τουρισμός για Όλους» 2013–2014 [Tourism for All 2013–2014 Programme]. Available at: www.koinonikostourismos.gr/pdfs/tourgiaolous2013_2014.pdf (accessed 21 November 2018).

ETHNOS (2020) Κοινωνικός Τουρισμός: Νέοι δικαιούχοι στο πρόγραμμα – Πότε λήγει η προθεσμία [Social Tourism: New beneficiaries – Programme's deadline]. Available at: https://www.ethnos.gr/oikonomia/114167_koinonikos-toyrismos-neoi-dikaioyhoi-sto-programma-pote-ligei-i-prothesmia (accessed 21 October 2020).

Eusébio, C., Carneiro, M.J., Kastenholz, E. and Alvelos, H. (2016) The impact of social tourism for seniors on the economic development of tourism destinations. *European Journal of Tourism Research* 12, 5–24.

Forster, E.S. (1958) *A Short History of Modern Greece, 1821–1956*, 3rd rev. edn. Methuen & Co. Ltd, London.

Goodin, E. (1988) *Reasons for Welfare: The Political Theory of the Welfare State*. Princeton University Press, Princeton, New Jersey.

Greek Government Gazette (1940) *Αναγκαστικός Νόμος 2436, 8 Ιουλίου 1940 'Περί συστάσεως τουριστικού δελτίου' [Mandatory Law No. 2436 of 8 July 1940 'Regarding the establishment of tourist coupon']*. National Printing Office, Athens.

Greek Government Gazette (1995) *Προεδρικό Διάταγμα Υπ' Αριθ. 225 [Presidential Decree No. 225], 20 June*. National Printing Office, Athens.

Greek Government Gazette (1996) *Προεδρικό Διάταγμα Υπ' Αριθ. 27 [Presidential Decree No. 27], 1 February*. National Printing Office, Athens.

Greek Government Gazette (2001) *Προεδρικό Διάταγμα Υπ' Αριθ. 343 [Presidential Decree No. 343], 11 October*. National Printing Office, Athens.

Greek Government Gazette (2002) *Νόμος Υπ' Αριθ. 3050 [Law No. 3050], 13 September*. National Printing Office, Athens.

Greek Government Gazette (2009) *Προεδρικό Διάταγμα Υπ' Αριθ. 186 [Presidential Decree No. 186], 7 October*. National Printing Office, Athens.

Greek Government Gazette (2014) *Προεδρικό Διάταγμα Υπ' Αριθ. 112 [Presidential Decree No. 112], 29 August*. National Printing Office, Athens.

Greek Government Gazette (2015) *Προεδρικό Διάταγμα Υπ' Αριθ. 70 [Presidential Decree No. 70], 22 September*. National Printing Office, Athens.

Greek Government Gazette (2016) *Προεδρικό Διάταγμα Υπ' Αριθ. 123 [Presidential Decree No. 123], 4 November*. National Printing Office, Athens.

Greek Government Gazette (2018) *Προεδρικό Διάταγμα Υπ' Αριθ. 72 [Presidential Decree No. 72], 2 August*. National Printing Office, Athens.

Greek Government (2020) Πρόγραμμα ενίσχυσης της ζήτησης του εγχώριου τουρισμού «Τουρισμός για όλους» έτους 2020 [2020 'Tourism for All' programme to boost domestic tourism demand]. Available at: https://covid19.gov.gr/programma-enischysis-tis-zitisis-tou-egchoriou-tourismou-tourismos-gia-olous-etous-2020/ (accessed 21 October 2020).

Haukeland, J.V. (1990) Non-travellers: The flip side of motivation. *Annals of Tourism Research* 17(2), 172–184.

Igoumenakis, N.G., Kravaritis, K.N. and Lytras, P.N. (1998) *Εισαγωγή στον Τουρισμό [Introduction to Tourism]*. Interbooks, Athens.

Kakoudakis, K.I. (2020) Η χρησιμότητα του κοινωνικού τουρισμού σε περιόδους οικονομικής ύφεσης [The usefulness of social tourism in times of economic recession]. HuffPost Greece. Available at: https://www.huffingtonpost.gr/entry/e-chresimoteta-toe-koinonikoe-toerismoe-se-periodoes-oikonomikes-efeses_gr_5eb265cbc5b6f8deb4e97b0d (accessed 22 October 2020).

Kakoudakis, K.I. and McCabe, S. (2018) Social tourism as a modest, yet sustainable, development strategy: policy recommendations for Greece. *Journal of Policy Research in Tourism, Leisure and Events* 10(3), 189–203. Available at: https://doi.org/10.1080/19407963.2018.1443938 DOI: 10.1080/19407963.2018.1443938.

Kakoudakis, K.I., McCabe, S. and Story, V. (2017) Social tourism and self-efficacy: exploring links between tourism participation, job-seeking and unemployment. *Annals of Tourism Research* 65, 108–121. Available at: https://doi.org/10.1016/j.annals.2017.05.005

Kathimerini (2012) «Λουκέτο» σε ξενοδοχεία φέρνει η κατάργηση των προγραμμάτων κοινωνικού τουρισμού [The abolishment of social tourism programmes results in hotel shutdowns]. Available at: www.kathimerini.gr/451515/article/oikonomia/epixeirhseis/loyketo-se-3enodoxeia-fernei-h-katarghsh-twn-programmatwn-koinwnikoy-toyrismoy (accessed 7 December 2018).

Katsigiannis, K.C. (2017) *Ελληνικός Οργανισμός Τουρισμού: Ταξίδι στο χρόνο [Greek National Tourism Organization: Journey in Time]*. Mitropolis, Athens.

KEPE (1987) *Programming Issues E9. Reports for the 1983–1987 Programme: Tourism*. KEPE, Athens.

Klitgaard, M.B. (2007) Why are they doing it? Social democracy and market-oriented welfare state reforms. *West European Politics* 30(1), 172–194. DOI: 10.1080/01402380601019753.

La Placa, V. and Corlyon, J. (2014) Social tourism and organised capitalism: research, policy and practice. *Journal of Policy Research in Tourism, Leisure and Events* 6(1), 66–79. DOI: 10.1080/19407963.2013.833934.

Logothetis, M.I. (1967) Προβλήματα και κατευθύνσεις τουριστικής πολιτικής [Problems and directions of tourism policy]. Unpublished PhD thesis. Panteion University of Social and Political Sciences, Athens, Greece.

Lowe, R. (1993) *The Welfare State in Britain Since 1945*. Macmillan, Houndmills and London.

Matsaganis, M. (2011) The welfare state and the crisis: the case of Greece. *Journal of European Social Policy* 21(5), 501–512. DOI: 10.1177/0958928711418858.

Miller, J.E. (2009) *The United States and the Making of Modern Greece: History and Power, 1950–1974*. University of North Carolina Press, Chapel Hill, North Carolina.

Minnaert, L., Maitland, R. and Miller, G. (2006) Social tourism and its ethical foundations. *Tourism Culture & Communication* 7(1), 7–17. DOI: 10.3727/109830406778493533.

Minnaert, L., Diekmann, A. and McCabe, S. (2012) Defining social tourism and its historical context. In: McCabe, S., Minnaert, L. and Diekmann, A. (eds) *Social Tourism in Europe: Theory and Practice*. Channel View Publications, Bristol, UK, pp. 18–30.

Mishra, R. (1981) *Society and Social Policy, Theories and Practice of Welfare*, 2nd edn. Macmillan, London.

Nikolakakis, M. (2013) Τουρισμός και ελληνική κοινωνία την περίοδο 1945–1974 [Tourism and Greek society during the period 1945–1974]. Unpublished PhD thesis. University of Crete, Heraklion, Greece.

OAED (2018a) Δημόσια Πρόσκληση 05/2018. Πρόγραμμα επιδότησης διακοπών εργαζομένων, ανέργων και των οικογενειών αυτών με επιταγή κοινωνικού τουρισμού περιόδου 2018–2019 [Public call 05/2018. 2018–2019 programme of subsidised holidays for employed and unemployed individuals, and their families, through social tourism coupons]. Available at: www.oaed.gr/koinonikos-tourismos-2018-2019 (accessed 9 September 2019).

OAED (2018b) Δημόσια Πρόσκληση 03/2018. Πρόγραμμα διαμονής παιδιών σε παιδικές κατασκηνώσεις έτους 2018 [Public call 03/2018. 2018 Camp for kids programme]. Available at: www.oaed.gr/kataskenotiko-programma (accessed 9 September 2019).

OAED (2020) Δελτίο Τύπου – Εναρξη υποβολής αιτήσεων [Press release – Start of applications' submission]. Available at: www.oaed.gr/koinonikos-tourismos-2020-2021 (accessed 21 October 2020).

OAED (n.d) ΟΑΕΔ: Ιστορική Εξέλιξη [OAED: Historical progress]. Available at: www.oaed.gr/istorike-exelixe (accessed 23 June 2019).

OEE (2012) Ανακοίνωση του Πανελληνίου Συλλόγου Εργαζομένων Εργατικής Εστίας [Announcement of the Panhellenic Association of Ergatiki Estia's employees]. Available at: www.oee.gr/ (accessed 10 November 2019).

OEE (n.d) Κοινωνικός Τουρισμός [Social tourism]. Available at: www.oee.gr/koinonikos_tourismos.html (accessed 7 December 2018).

OGA (2017) Απολογισμός Διοίκησης Έτους 2016 – Επιχειρησιακό Σχέδιο Δράσης Έτους 2017. [2016 Administration Review – 2017 Operational Action Plan]. Available at: www.oga.gr/# (accessed 23 November 2018).

Oikonomou, C. (1996) Το κοινωνικό κράτος στη δεκαετία του 1980: Το παράδειγμα του Εθνικού Συστήματος Υγείας [The welfare state in the 1980s: The case of the National Healthcare System]. Unpublished PhD thesis. Panteion University of Social and Political Sciences, Athens, Greece.

OPEKA (2018) Πρόγραμμα Αγροτικής Εστίας έτους 2018 [Agrotiki Estia 2018 Programme]. Available at: www.koinonikostourismos.gr/pdfs/pliroforiesAgr_Estia_2018.pdf (accessed 17 September 2019).

Pagoulatos, G. (2003) *Greece's New Political Economy: State, Finance, and Growth from Postwar to EMU*. Palgrave Macmillan, London/New York.

Papadoulaki, A. (2011) Ιστορία των Τουριστικών Επιχειρήσεων στο 19ο και 20ό αιώνα. Ο ρόλος και η εξέλιξη των ταξιδιωτικών γραφείων & πρακτορείων [History of tourist enterprises in the 19th and 20th century. The role and development of tourist bureaus and agencies]. Unpublished PhD thesis. Hellenic Open University, Patras, Greece.

Raptarchis, P.K. (2007) Διαρκής Κώδικας Νομοθεσίας: Τόμος 18 – Τύπος και Τουρισμός [Abiding Law Code: Vol. 18 – Press and Tourism]. Hellenic Republic, Ministry of Interior. Available at: www.e-themis.gov.gr/Portal/Default.aspx?Page=downloads (accessed 23 June 2019).

Richards, G. (1998) Time for a holiday? Social rights and international tourism consumption. *Time and Society* 7, 145–160.

Sotiriadis, M. (1994) *Τουριστική Πολιτική [Tourist Policy]*. Technological Educational Institute of Crete, Heraklion, Crete.

Tourism Issues (1983) *Κοινωνικός Τουρισμός [Social Tourism]*, pp. 46–.47.

Tourismos & Economia (1974) *Εσωτερικός Τουρισμός [Domestic Tourism]*, pp. 42–.45.

Tourismos & Economia (1986) *Συνέδριο του Διεθνούς Γραφείου Κοινωνικού Τουρισμού [International Bureau of Social Tourism Conference]*, p. 40.

Tsakalotos, E. (1998) The political economy of social democratic economic policies: The Pasok experiment in Greece. *Oxford Review of Economic Policy* 14(1), 114–138. DOI: 10.1093/oxrep/14.1.114.

TYPET (2018) Camps. Available at: http://typet.gr/index.php/en/camp (accessed 9 September 2019).

Vlachos, A.P. (2013) Τουριστική ανάπτυξη και δημόσιες πολιτικές στη σύγχρονη Ελλάδα, 1914–1950: Η ανάδυση ενός νεοτερικού φαινομένου [Tourism development and public policies in contemporary Greece, 1914–1950: The emergence of a modernist phenomenon]. PhD thesis. National and Kapodistrian University of Athens, Athens, Greece.

Walton, J.K. (2013) 'Social tourism' in Britain: history and prospects. *Journal of Policy Research in Tourism, Leisure and Events* 5(1), 46–61. DOI: 10.1080/19407963.2012.703377.

2 Social Tourism in Mexico: Evolution, Challenges and Future Development

Adriana E. Estrada-González*
School of Business, Universidad de Monterrey, Mexico

2.1 Introduction

The right to annual paid vacations has been considered an initial condition for social tourism development (Cestur-UAM, 2000). Travel for leisure was an ambition of manufacturing workers and their families during the 19th century, as well as an idealized goal for many other workers. Despite the Geneva Convention statement, adopted on 26 June 1936, which stated that employees should be granted paid vacations, several countries saw social tourism as an avant-garde trend and incorporated it into their national constitutions and bylaws. However, in the mid to late 19th century during the latter part of the Industrial Revolution, mass movements (and mass tourism) became more common when weekly working hours decreased and the middle class could afford to take time off (Bardolet, 2016).

By the 20th century, the general granting of paid vacations to employees led to a significant expansion in tourism. Known as the 'Holidays with Pay Recommendation' (No. 47, 1936), the General Conference of the International Labour Organization convened in Geneva adopted certain proposals regarding annual holidays with pay for employed people (ILO, 1936) to allow them to rest, have recreation and develop their faculties.

Several years after ratifying the Geneva Convention, human rights became an international concern. While some countries were observing human rights, many more were committing severe violations. As a result, unified criteria for protecting human rights were proposed, with representatives from different legal and cultural backgrounds gathering to draft a declaration. In 1948, the United Nations General Assembly proclaimed the Universal Declaration of Human Rights, which outlined the need for leisure/free time and the access to vacations in its 24th Article (UN, 1948): 'Everyone has the right to rest and leisure, including a reasonable limitation of working hours and periodic holidays with pay.' Following this international right, several governmental models, societal initiatives, private enterprise programmes and non-government organizations (NGOs) agreed on the common idea to grant the right to leisure to everyone in society. As a result, programmes, initiatives, the right to paid holidays and several actions emerged such as social tourism, which has prevailed for the past 60 years.

Norberto Tonini, the former chairman of the International Social Tourism Organisation (ISTO), declared that social tourism was a solidarity function to provide access to leisure vacations to groups who were disadvantaged because of low wages or other economic constraints (Tonini, 2006). However, people do

*adriana.estrada@udem.edu

 Social Tourism: Global Challenges and Approaches (eds J. Lima and C. Eusébio)
DOI: 10.1079/9781789241211.0003

not actively participate in tourist activities for several reasons, such as health, family obligations or work engagements. Further analysis is necessary to understand the complex environment surrounding social tourism. This chapter aims to contribute to this analysis, reflecting on Mexico's reality.

This chapter firstly provides a discussion regarding the definition of social tourism, followed by an analysis of social tourism in Mexico and its development in the past 45 years. The chapter concludes by presenting policies and initiatives related to social tourism.

2.2 Defining Social Tourism

The common background to social tourism arose from the different organizations involved at the time. In 1962, several national and international associations gathered in a Congress in Milan and decided to establish an international organization to deal with social tourism, which gave rise to the establishment of the Bureau International du Tourisme Social (BITS) in 1963. In 2010, BITS changed its name to the ISTO (International Social Tourism Organisation), with the head office remaining in Belgium; however, since its establishment, the organization has been extremely active in publishing international declarations and position papers. Among its statutes, Article 3 gave a wide definition for social tourism (Bureau International du Tourisme Social (BITS), 1963):

> Social tourism is understood as the set of relations and phenomena involved in tourism participation, particularly by people with low incomes. This participation is possible through well-defined and applied social strategies.

Since this definition was conceptualized 55 years ago, there have been several updates over the years, mainly through working papers and declarations, such as the Vienna Charter (1972), the Montreal Declaration (1996) and the Aubagne Addendum (2006) (BITS, 1972, 1996, 2006).

In the Vienna Charter, BITS gave an operational definition rather than the visionary goal stated originally in its 1963 Statutes. This definition was revised with the focus on 'social tourism as programmes, initiatives and actions granting the right to vacations and tourism accessibility to all groups of population, especially for youth, families, retired citizens, handicapped people, and low-income individuals' (Jolin, 2003).

In 1996, BITS held its biannual congress and adopted a new social tourism concept in a new document known as the Montreal Declaration, wherein the challenges associated with exclusion and integration were recognized. At the core of this new definition was recognition of the struggle to overcome the inequalities and exclusion of people with low wages, the physically handicapped, those living in underdeveloped regions and those from different cultures (BITS, 1996).

The last declaration updated by ISTO was the Aubagne Addendum. A working committee was asked to review the Montreal Declaration articles and write an addendum with ten updated propositions. Proposition 5 supported the original definition, but focused on 'providing special attention to families, youth, and elderly people, and the development of trips for children and youth, including school field trips. These trips were relevant to discovering new environments and people and aiding social cohesion as well as experiencing a feeling for travelling' (BITS, 2006). The simplified social tourism definition used today by ISTO is any activity that contributes to greater access to holidays and tourism activities for everyone (ISTO, 2017).

Moreover, other organizations have defined social tourism. The World Tourism Organization (WTO) is a worldwide association charged with the promotion of tourism development. In September 1980, during its World Tourism Conference in Manila, Philippines, the organization proposed the Manila Declaration on World Tourism, which saw social tourism as something which 'a society must provide to all citizens less privileged to exercise their right to rest' (WTO, 1980). As such, it defined social tourism as 'making tourist leisure accessible to the majority, including youth, families and the elderly, by providing them with an exceptional economic opportunity'.

The Family Holiday Association is a British charity that has been operating for over 40 years to provide short trips to families struggling with issues such as disabilities, severe and sudden illnesses and scarce resources (FHA, 2018). This charity believes that holidays are a lifeline, not a luxury, which is their main reason for favouring family holidays, and define social tourism as 'the inclusion of groups in society who are economically weak or otherwise disadvantaged in tourism participation' Family Holiday Association (FHA) (n.d.).

There have been several academics who have also proposed a definition for social tourism. Muñiz (2001, p. 43) identified social tourism as 'activities generating tourism demand characterised by low economic resources, where access to leisure tourism can only be performed through the intervention of tour operators providing collective benefits to social tourists'. Another definition is proposed by Lominé and Edmunds (2007, pp. 170–171), who understand social tourism as 'tourism oriented by a political and psychological belief that individual well-being is acquired by travelling and being on vacation, benefitting the whole society through personal development and social cohesion'.

A recent study concerning advances in research on social tourism reviewed its definition as 'all activities, relationships and phenomena in the field of tourism resulting from the inclusion of otherwise disadvantaged and excluded groups in participation in tourism' (Diekmann *et al.*, 2018, p. 182).

As indicated, several social tourism definitions, both from the sector and from academics, have been proposed; however, in general, they all share the same spirit: promoting access to travel and tourism for everyone. It is evident that social tourism has been a concern in many countries, and Mexico is no exception. The following section deals with Mexican tourism development, the introduction of social tourism and the state policies, regulations and management of social tourism.

2.3 The Development of Social Tourism in Mexico

Social tourism development in Mexico is relatively recent as it was only during the 20th century that tourism became institutionalized and involved public, private and social participation. Mexican governments have assisted in developing the major tourist destinations, and the private companies (tour operators, the hospitality industry and tourist attractions) have supplied tourist services for both domestic and international audiences. However, it was only during the 1980s that Mexico decided to establish some social tourism programmes. In Europe, several countries had begun developing social tourism policies, programmes and working plans, among which was France, with its very early 'Tourism for All' and right for leisure initiatives for manufacturing workers (Muñiz, 2001). France approved paid vacations in 1936, followed by an increase in vacation days and the 35-hour working week. Consequently, this French model was adopted in Mexico, after which specific public policies were developed. In the beginning, the Mexican model provided government funding to support citizens who were unable to use their own resources to travel. After a few years, the focus of the social tourism programmes changed, with the emphasis refocusing on collaborative private and social partnerships. NGOs focused on tourism then opened to fill the gap left by the declining government support and the effect of the economic recession on private enterprises.

This Mexican 'new tourism' structure was unique as social tourism was part of domestic tourism and was integrated into leisure and recreation travel but was offered to citizens with low incomes. To meet their travel requirements, institutional sponsorships were offered and tourist providers offered low-cost trips to the people who were unable to pay for their own accommodation, transportation and recreation needs (Cestur-UAM, 2000).

After World War II, there was a worldwide boom in tourist organizations providing vacation programmes to marginalized groups (Cazes, 2000), mainly owing to economic constraints and because of the various leisure travel government assistance plans and state subsidies. The tourism industry was growing rapidly and regulations were established to standardize the quality of both public and private tourist service agencies.

Social tourism policies in Mexico have had a clear statement over the years: 'Tourism participation must be accessible to every Mexican' (Cestur-UAM, 2000, p. 5). However, this statement entailed state intervention, whereby every social group, through specific subsidies, could have access to tourism, regardless of their economic status and regular income (Cestur-UAM, 2000, p. 5). Believing that state intervention would solve major obstacles for tourism participation, the Federal Tourism Law included a Social Tourism Chapter. The relative economic importance of the tourist industry and its contribution to the national GDP finally led to the establishment of a separate Ministry of Tourism in 1974, including social tourism programmes and its chapter into the law.

Since the introduction of a Federal Tourism Law in 1949 to Mexican public policies, the amended 1974 law included four articles addressing social tourism. The approach of these four articles involved developing plans and programmes to provide leisure and tourism participation as a strategy for accessibility for every Mexican citizen, regardless of their economic condition. The government tried to launch a social tourism project where resources were provided exclusively by public funds, akin to a welfare state. Article 35, from the referred law, stated that the 'National Tourism Department will promote social tourism, where every Mexican has access to tourism programmes' (DOF, 1974). In this capacity, the government was in charge of sealing agreements with public and private companies in order for every citizen to travel within the national boundaries through family or group tour packages at reduced prices. A few national organizations were approached, mainly those gathering youths, senior citizens or state workers.

To realize a welfare state, public subsidies to support Mexican citizens to travel at reduced rates were established for the first time in history. Over the past 75 years, government actions to support underprivileged groups through subsidies had been mainly for immediate needs such as food products (corn, beans, milk, rice, eggs), dwelling or public transportation; tourism services were then included among these public subsidies (Estrada-González, 2016).

The remaining articles in this first Social Tourism Chapter mentioned tourism suppliers' participation (tour operators, travel agencies) and how public tourism programmes should be planned regarding tourism seasonality, folk and traditional events, and Mexican festivities (DOF, 1974). Particular attention was paid to the groups of citizens to be supported by public subsidies: students, civil servants, teachers, general workers and small landowners.

By 1980, a new law amendment had been signed. In general, the law was strengthened, as was the Social Tourism Chapter, growing to 14 articles from the original four. The amended chapter reflects a civic and patriotic spirit. This patriotic spirit was written in an article as 'to emphasise national identity and solidarity, through the knowledge and proper use of national tourism heritage' (DOF, 1980). It also proposed creating cooperative organisations and incorporating goods, materials and equipment for public and social suppliers. A substantial change in this chapter was the inclusion of public and private tourism suppliers, to support tourism provision at reduced rates, seeking alternative resources to provide tourism participation. Looking for private suppliers' participation, the government dealt with new agreements in the form of discounts or financing credit for tourism services.

Regarding the population groups supported by the law, the youth market was included in Article 85, as was the need to focus on cultural, athletic and civic events during vacation seasons. These events had to be planned during vacation periods to make a rational use of national resources and tourism infrastructure (DOF, 1980). Clearly, the government decided to solve the seasonality problem by offering reduced rates during low seasons.

These actions and challenges laid the groundwork for the establishment of social tourism. Some national researchers viewed social tourism as a fashionable tourism trend, academics saw social tourism as a tourism product and tourism providers saw it as a way to stabilize seasonality. Although each of these propositions correctly reflected the interests of each different stakeholder, social tourism proved to be a remarkable tool for Mexican citizens with limited financial resources to spend on leisure travel. The section on social tourism policies exemplifies the amendments to the law.

2.4 Social Tourism Programmes in Mexico

The transitions experienced from the 19th to the 20th century increased people's motivations to travel, with tourism being viewed as giving people the opportunity to enjoy new experiences and pleasure, avoid everyday chores and escape from daily routines (Rodríguez, 1982).

These tourist experiences promoted a sense of inclusion in society, with working families who had few resources to travel and enjoy leisure tourism seeking financial support to travel from both public and private agencies. An unexpected expansion of tourist flow emerged, and as a result, the Mexican government decided to further closely supervise tourism suppliers to prevent inappropriate behaviour.

Boullón (1983) classified tourism into three categories: foreign, outbound and domestic (including commercial, social and sponsored tourism). The two different social tourism and sponsored tourism sub-categories in Boullón's classification for domestic tourism have sometimes been misunderstood. Social tourism, however, was usually supported by a state agency – Turissste, created in 1972, or labour unions, while sponsored tourism included assistance by private companies or social associations to specific groups such as disadvantaged families, senior citizens and youth.

When social tourism was included in the Federal Tourism Law in 1974, there were several actions undertaken by the government. The Tourism Department first identified the different population groups that were not a part of the economically active population, with workers, peasants, students, civil servants and retired citizens being the first groups to receive support to travel and participate in tourism (Gobierno del Estado de México, 1981). The state offered camping facilities, family picnics over weekends and family hostelling in summertime. By 1977, Desarrollo Integral de la Familia (DIF), a government institution, provided a 10-day family camping programme that included ecological awareness and recreational, educational and cultural activities.

By the early 1980s, social tourism programmes began to lose relevance in the public agencies, and there was a significant decrease in tourism provisions. The worldwide financial crisis had left state programmes without investors, which had a flow-on effect on social tourism provision. As a result, government departments began to look for new tourism providers and approached private companies. In 1986, the Immediate Action Plan for Tourism Promotion outlined new plans that included supporting road travel and promoting low-price tour packages that included airfares (Torruco Marqués, 1988), all of which resulted in an increase in 'new passengers' travelling to tourist destinations. Charter flights and tour packages became viable options for the working classes, and many manufacturing workers, students and retired persons were able to fly for the first time in their lives. Social tourism emerged once again, but it was now part of the private tourism industry.

During the 1990s, private organizations and NGOs, such as the National Institute for Senior Citizens, Unidos (an association for disabled youth) and religious charities were seeking to restore social tourism. Tour packages had proven to be a successful format for balancing seasonality in tourism destinations, with many tourism providers offering attractive discounts in the low season. Activities were reoriented to promote ecotourism, sports and recreational activities, culture and health tourism. The French social tourism model was revised again and new values were added to envision the future: accessibility, solidarity and sustainability, with a focus on community development rather than mere revenue-raising.

By the turn of the century, the international trend to rename some tourism programmes flowed on to social tourism and a global trend emerged known as 'Tourism for All'. In 1990, a Tourism for All trade show took place at which tourism infrastructure was the most popular feature especially for families who could not afford to go on holidays (Sectur, 2000). Several initiatives followed the trade show such as the creation of 'Vacaciones Inolvidables' (unforgettable vacations) – a programme designed by DIF, the distribution of brochures and magazines promoting the state travel agency and seaside resorts and publications such as the Low Cost Accommodation Directory, the Hostels and Camping Directory and the Road Vehicle Parks Directory.

The Ministry of Tourism developed a programme focused on Mexicans living in the United States, and particularly those living in Texas, New York, Illinois, California, Arizona and Florida. Between 1989 and 1997, many Mexicans living abroad were returning to Mexico for their holidays and to visit family and friends, which resulted in a 73.5% increase in 'tourist arrivals' and with social tourism playing a relevant role in offering appropriate services for returning Mexicans (BITS, 1999). This programme was similar to one that had been

Table 2.1. Social tourism initiatives and programmes in Mexico.

1972	Creation of Turissste, a state travel agency devoted to promote social tourism
1974	Inclusion of the Social Tourism Chapter in Federal Tourism Law
1977	Camping facilities available for family tourism by Family Development (DIF)
1980	Decrease of public support for social tourism programmes
1986	Establishment of the Immediate Action Plan for Tourism Promotion
1990s	Promotion of social tourism among seniors (National Institute for Senior Citizens)
1989–1997	Social tourism programme for Mexicans living abroad
1990	Tourism for All trade show

implemented in Spain during the 1990s, organized by the Ministry of Labour and Social Affairs. Holidays for Seniors was a programme for senior citizens above the age of 65, pre-retired people below the age of 60 and widows above the age of 55. It especially focused on Spanish immigrants residing in European countries who had not been able to return to Spain for 40 years, mainly because of economic constraints (BITS, 1999). Table 2.1 summarizes the evolution of initiatives and programmes developed by state agencies supporting social tourism. The next section describes the policies, plans and best practices associated with social tourism in Mexico.

2.5 Social Tourism Policy in Mexico

The evolution of the Federal Tourism Law and its implications on social tourism development have been previously discussed. The 1974 update incorporated social tourism into law and since that time, there have been substantial changes in the government structure. Some relevant public policies and initiatives that have addressed social tourism are: (i) the Federal Tourism Law and its associated regulations; (ii) the Social Tourism National Council and its associated regulations; and (iii) the Agreement to establish a Tourism for All Public Policy.

2.5.1 Federal Tourism Law

Social tourism was incorporated into the law as a relevant programme in 1974, and the main goal was to provide an accessible tourist plan to meet current demand (DOF, 1974). Similarly to the adapted model from European countries, this first development stage was solely a government plan, with only government resources being used to promote the social tourism programmes. Public agencies actively participated and signed agreements with tourism providers to facilitate domestic travel for citizens using family or group packages at reduced prices. These agreements were focused particularly on travel for senior citizens, young people and civil servants (DOF, 1974), and the social tourism programme was oriented to specific market segments such as students, teachers, civil servants, workers and small business owners.

The updated law in 1980 reflected the main goal to 'stimulate national identity and solidarity through the knowledge and use of national tourist heritage resources and promote participation by co-operative associations to provide goods and services' (DOF, 1980). Therefore, there was a serious focus on inclusion in this law as the government was hoping that every national citizen would be able to participate in the tourist activities; consequently, youth were added as a new focused market segment. At the same time, programmes were being designed to emphasize civic, sports and cultural events, which were organized during the vacation season to ensure a rational use of the tourist resources and infrastructure.

New legal requirements were then incorporated to promote and stimulate the use of social tourism facilities and encourage programme investment, with the government also offering financial support to associations, cooperative groups and committees to provide social tourism services. One distinctive feature of these groups was the creation of savings programmes to promote social tourism for financially deprived citizens. Another new legal section was required to allow for the conduct of research and surveys to gather the statistics necessary to update the planning, promotion and operating mechanisms. This work was done by tourist providers, investors, training centres, universities and research institutes, and after the data and statistics had been collected, there were seminars, congresses and meetings organized at regional, national and international levels (DOF, 1980).

In 1984, the Federal Law amendment was rather conservative. Nonetheless, all regulations included in the previous iteration remained with only some minor changes; for example, children and non-salaried workers (workers who do not work regularly for a company, and hence do not have a regular salary) were added to the market segments already supported and specific interest was placed on promoting natural, historic, cultural and typical tourist attractions. However, some regulations were also added, such as agreements between tourist providers and social tourists to offer discounts and reduced prices, investment stimuli to increase tourist facilities and international social tourism exchange programmes (Ley Federal de Turismo, 1984).

At the turn of the 20th century, social tourism had evolved and so had its interpretation; therefore, the law was updated with a revised social tourism definition (DOF, 1992, p. 3).

> Social tourism includes those means and instruments, through which people with limited financial resources are supported to travel in adequate economical, secure and comfortable conditions.

In addition to this updated definition, the law outlined the public agencies that were to be involved in providing tourist services: the Ministry for Social Development, the Social Security and Health Institute for State Employees, the Mexican Social Security Institute, the National Sports Commission and the Ministry of Tourism (DOF, 1992).

The amendment to the 1999 Federal Law was made to provide adequate social tourism services to disabled people, which was a new market segment. Therefore, the legal amendments were aimed at providing more services to include the disabled and to ensure accessibility (DOF, 1999). The 2009 amendment maintained the same scope, with the main market segments being families, senior citizens, youth and the disabled, and the only visible changes being in the vocabulary; for example, changing the term 'elderly' to 'senior citizens' and the term 'disabled' to people with 'diverse capabilities' (DOF, 2009).

In 2015, the law experienced its most recent amendment (DOF, 2015). The social tourism provision remains unchanged but a principle of equality was included in the chapter. With this in mind, the Ministry of Tourism proposed that every Mexican has the same right to access to social tourism. Furthermore, the law included the National Council for Sports and Physical Culture to participate in the development of social tourism programmes, including sports and fitness. Table 2.2 summarizes the relevance of each amended law.

Table 2.2. Relevant changes in the Social Tourism Chapter (General Tourism Law).

Tourism law	Amendments
1974	Social tourism chapter inclusion in the law. Social tourism is a public programme oriented to students, teachers and civil servants
1980	Intention to stimulate national identity and solidarity through the knowledge and use of national tourist heritage resources during social tourism tours
1984	Children and non-salaried workers were included as groups to be supported by social tourism programmes
1992	An updated social tourism definition was included in the law
1999	Disabled people were included as a group to receive financial support to travel
2009	Changes in the vocabulary of supported groups: from 'elderly' to 'senior citizens'; from 'disabled' to people with 'diverse capabilities'
2015	Equality is observed to provide social tourism services to every Mexican

2.5.2 Social Tourism National Council

There had been efforts to promote social tourism in the Mexican population from its inception into the Federal Tourism Law, with financial programmes for entrepreneurs and small companies being designed to support social tourism development. However, many of these efforts appeared to be disconnected as many companies, state agencies and social organizations were seeking to accomplish the same things. With the development of the Integral Development Programme for Social Tourism (2001–2006), the Social Tourism National Council was established with the goal of developing and consolidating tourist facilities/infrastructure, stimulating tourism demand, and promoting social tourism products and services. Consequently, several states in Mexico established their own Social Tourism State Councils (Sectur, 2006) in which different initiatives were proposed based on the state environment and with the objectives to provide valuable recreational programmes for local residents, promote social tourism in their tourist destinations and contribute to local development. These state councils were multinetwork entities in which public, private and social organizations collaborated. Among others, government employees, tour operators, hotel and restaurant associations, universities and NGOs were the main actors working together to provide appropriate social tourism services to tourists eager to fully participate.

The tourist services offered by the public, private and social organizations are tour packages including transportation and accommodation, day tour activities to aquatic parks, swimming pools and recreation centres, tours along historic and architectural routes, beach resort developments or stays in bungalows and rustic cabins in the mountains, all of which are offered to families, senior and retired citizens, and children. In particular, the children's programmes are designed so that they can experience different environments; that is, rural children are taken to visit city landscapes and urban children are taken to rural areas.

Benefits derived from participation in social tourism are increased tourist flows during the low season, domestic tourism stimulus, destination diversity, competitiveness, and an increase in productivity and employment. In addition, social tourism has been found to promote tourism supply and contribute to enterprise profitability.

One of the public agencies that is part of the National Council is Turissste, which is the travel agency that offers services to government employees and is part of the Social Security and Health Institute for State Employees. Another public agency is the Mexican Social Security Institute (IMSS) offering four vacation centres that provide recreational activities and lodging to affiliated employees: the National Institute for Senior Citizens (INAPAM), which offers tourist activities at accessible prices; the Youth Mexican Institute, which offers travel discounts to 12–29 year olds; and the Anthropology and History National Institute (INAH), which is in charge of the archeological sites, cultural and historic heritage sites, and artistic and historic monuments. All these agencies provide unique programmes to promote social tourism to the Mexican citizenry (Table 2.3).

Table 2.3. Public agencies initiatives on social tourism.

Council's agencies	Initiatives
Turissste	Provides travel services for state employees, offering social tourism programmes at designated tourism resorts and destinations
IMSS	Provides accommodation and recreation services at their social tourism vacation centres year-round
INAPAM	Provides social tourism day trips and vacations to senior citizens
INAH	Promotes national heritage sites, including them in itineraries and social tourism travel packages

2.5.3 Agreement to establish a Tourism for All Public Policy

In 2006, approval was given to create a Tourism for All Public Policy, with the goal of developing social tourism opportunities through collaborative public, private and social agency efforts. This policy commits the Mexican government to providing opportunities for families and individuals to access social services to promote well-being in order to develop a more inclusive, responsible and participative society (Diario Oficial de la Federación (DOF), 2006).

The Tourism for All Public Policy promotes four major features to achieve general well-being: social inclusion, leisure, recreation and free time. It has two main objectives: to promote the desire for travel and tourism among Mexicans and to seek actions that are aligned with public policies to emphasize the promotion, development and provision of tourism, recreation and free time facilities. As the Tourism for All policy guarantees that tourism activities are available to every Mexican and are of adequate quality, a major focus of the policy is the development of the tourist goods and services required to provide the necessary tenets of social tourism: access to vacations, places for natural and cultural heritage enjoyment and contemplation, and appropriate infrastructure and services.

The Ministry of Tourism recognizes that there is a need to educate for leisure as many Mexicans are used to working long hours without regular vacations. The Ministry promotes the tourism and travel culture as having a right to rest and having creative periods to allow for personal development through leisure and tourism. However, it has taken a long time to change citizens' mentality regarding tourism destinations: that they are not only for foreign tourists but also for Mexicans. Therefore, a major focus over the past 20 years has been to create a common understanding that everybody is valuable and has the same rights to enjoy leisure and tourism.

2.6 Conclusion

Tourism is a major modern leisure activity, in which public, private and social agencies are involved. Initial steps to shape a tourism regulation framework were undertaken by the Mexican government in 1949, which was followed by the establishment of federal tourism laws, public policies and programmes, and private tourist developments, all of which were conducive to a fast-growing tourism scenario in Mexico. As domestic tourism initiatives promoted alternative destinations far from the crowded beach and metropolis destinations, diverse tourism programmes emerged focused on promoting and developing different areas (colonial cities, ecotourism resorts, archeological sites). By the 1960s, Mexico started experiencing the beginning of 'mass tourism', which led to significant infrastructure and tourism services developments to cater for both domestic and foreign tourists.

By the mid-1980s, Mexico had focused its attention on new plans and programmes, mainly derived from European examples. Social tourism was accepted by both public and private organizations, and initial actions were developed to provide tourist services to Mexicans unable to regularly participate in tourism. The federal government began to establish specific programmes for labour union workers, peasants, low-level manufacturing employees and cooperative workers. This time was known as the state welfare period because of the strong government participation in the provision of social tourism programmes and development of guiding public policies to promote tourism participation among the underprivileged. By the end of the 20th century, government spending contracted and was replaced by a public–private tourism provision model in which subsidies and sponsorships were offered by the tourism industry companies such as the national airline carriers, the tour operators, hotel and food and beverage companies, theme parks and attractions, and entertainment services. The mechanism for this mixed-service structure was led by the government, along with state departments and agencies seeking reduced prices and discounts from the private companies (hotel industry, tour operators and travel agencies, airlines), who gave attractive discounts to the market segments supported by the government and had the right to select the dates for these discounts, which was usually during low season.

Social tourism in Mexico is an integral part of the domestic tourism market and is understood as 'any travel, with a leisure or recreation purpose, without the intention of profit' (Cestur-UAM, 2000, pp. 2–5). This kind of travel is offered to people with low incomes who need economic support to be involved in leisure tourism. The Federal Tourism Law states that social tourism provides a framework for the development of tourism programmes for special market segments, such as families, senior citizens, youth and the disabled.

There were some challenges in Mexico's tourism evolution and in the establishment of social tourism programmes, with the main challenge being planning continuity. The many changes in government departments, especially in the early tourism development (1929–1949) years, led to a series of disconnected actions and difficulty in developing coherent tourism regulations, with the creation and disappearance of commissions, councils and state departments all having a negative impact on tourism development. However, from 1949, with the adoption of the first Federal Tourism Law, the tourism industry began to become more organized, and after the establishment of the Ministry of Tourism in 1974, there has been further stable development.

Another challenge faced by social tourism has been the availability of financial resources. The economic support to travel for groups in need is provided by subsidies, sponsorships, travel discounts, reduced prices or financial support, which are covered by both social and private sources, or through agreements between public agencies, private companies or tourism providers. Although these agreements may take a long time to be established, once signed, they may last for a long time and benefit a large portion of the population.

There are possible future social tourism developments. First, the present market segments need to be reviewed to determine whether there is a need to incorporate new groups of citizens who might benefit from social tourism programmes. As families, seniors, disabled people and youth are currently supported by social and private organizations, it is time to redirect the government efforts to different population segments, such as single parent families, unemployed citizens or refugees and migrants.

Social tourism has been part of tourism research and practice for the past 60 years, and we believe that it will continue to be an excellent tool for promoting domestic tourism in Mexico. Although attracting foreign visitors is important for the local and national economies, social tourism provides the domestic market with the opportunity for leisure and tourism experiences. It is evident that there will be more changes to the programme formats, but regardless of whether the social tourism programmes remain stable or are adapted to new tourism demands, there will always be people interested in enrolling in the plans and programmes provided by different entities to support Mexican leisure travel and tourism.

References

Bardolet, E. (2016) International tourism. In: Jafari, J. and Xiao, H. (eds) *Encyclopedia of Tourism*. Springer International Publishing, Gland, Switzerland, pp. 484–486.

Boullón, R.C. (1983) *Tourism and Recreation Activities. Man as Protagonist. [Las actividades turísticas y recreativas. El hombre como protagonista]*. Editorial Trillas, S.A. de C.V, Mexico City.

Bureau International du Tourisme Social (BITS) (1963) *Statutes. [Statuts]*. BITS, Brussels.

Bureau International du Tourisme Social (BITS) (1972) *Vienna Charter. [Charte de Vienne]*. BITS, Brussels.

Bureau International du Tourisme Social (BITS) (1996) Montreal Declaration 'Towards a humanist, social vision of tourism'. [La Déclaration de Montréal 'Pour une vision humaniste et sociale du tourisme']. BITS, Brussels. Available at: https://bitsamericas.files.wordpress.com/2010/02/declaration-de-montreal-fr.pdf (accessed 3 September 2018).

Bureau International du Tourisme Social (BITS) (1999) *Social Tourism in the World. [Le Tourisme Social dans le Monde]*. No. 133. BITS, Brussels, pp. 5–10. Available at: http://issuu.com/oitsisto/docs/n__133

Bureau International du Tourisme Social (BITS) (2006) *Aubagne Addendum 'Towards a development and solidarity tourism'. [L'Addendum d'Aubagne 'Vers un tourisme de développement et de solidarité']*.

BITS, Brussels. Available at: https://bitsamericas.files.wordpress.com/2010/02/declaration-de-montreal-fr.pdf

Cazes, G. (2000) Social tourism. In: Jafari, J. (ed.) *Encyclopedia of Tourism*. Routledge, Abingdon.

Centro de Estudios Superiores en Turismo–Universidad Autónoma Metropolitana (2000) *Social Tourism viability survey in Mexico, Executive Report. [Estudio de viabilidad del turismo social en México, Informe Ejecutivo]*. Secretaría de Turismo, CESTUR-UAM, Mexico City.

Diario Oficial de la Federación (DOF) (1974) *Federal Law for Tourism Promotion. [Ley Federal de Fomento al Turismo]*. DOF, Mexico City.

Diario Oficial de la Federación (DOF) (1980) Tourism Federal Law. [Ley Federal del Turismo]. DOF, Mexico City. Available at: http://dof.gob.mx/nota_detalle.php?codigo=4845911&fecha=01/15/1980 (accessed 25 September 2018).

Diario Oficial de la Federación (DOF) (1992) Tourism Federal Law. [Ley Federal del Turismo]. DOF, Mexico City. Available at: www.ordenjuridico.gob.mx/Publicaciones/pdf/L113.pdf (accessed 25 September 2018).

Diario Oficial de la Federación (DOF) (1999) *Tourism Federal Law. [Ley Federal del Turismo]*. DOF, Mexico City.

Diario Oficial de la Federación (DOF) (2006) Agreement to establish a Tourism for All Public Policy. [Acuerdo por el que se establece la Política Pública de Turismo para Todos]. DOF, Mexico City. Available at: www.ordenjuridico.gob.mx/Federal/PE/APF/APC/SECTUR/Acuerdos/2006/30052006(1).pdf (accessed 25 September 2018).

Diario Oficial de la Federación (DOF) (2009) General Tourism Law. [Ley General de Turismo]. DOF, Mexico City. Available at: www.senado.gob.mx/comisiones/asuntos_migratorios/docs/legislacion/LG24.pdf (accessed 26 September 2018).

Diario Oficial de la Federación (DOF) (2015) *General Tourism Law. [Ley General de Turismo]*. DOF, Cámara de Diputados del H. Congreso de la Unión, Mexico City.

Diekmann, A., McCabe, S. and Ferreira, C.C. (2018) Social tourism: research advances, but stasis in policy. bridging the divide. *Journal of Policy Research in Tourism, Leisure and Events* 10(3), 181–188. DOI: 10.1080/19407963.2018.1490859.

Estrada-González, A.E. (2016) Social Tourism from a human leisure perspective. Senior travel in Nuevo Leon (Mexico). [El Turismo Social desde el Ocio Humanista. El caso de los viajes para personas mayores en el Estado de Nuevo León (México)]. PhD thesis. Universidad de Deusto, Deusto, Spain.

Family Holiday Association (FHA) (2018) Who we are. Available at: www.familyholidayassociation.org.uk/about-us/ (accessed 1 October 2018).

Family Holiday Association (FHA) (n.d.) *Everyone Needs a Holiday*. Assembly Package, FHA, London.

Gobierno del Estado de México (1981) *Social Tourism in Estado de Mexico. [Turismo Social en el Estado de México], Documentos de Turismo, No. 4, Dirección de Turismo 1975–1981*. Editora Lithomex, S. A, Mexico City.

International Labour Organization (ILO) (1936) R047–Holidays with pay recommendation, 1936 (NO. 47). Available at: www.ilo.org/dyn/normlex/en/f?p=NORMLEXPUB:12100:0::NO::P12100_ILO_CODE:R047 (accessed 19 September 2018).

International Social Tourism Organisation (ISTO) (2017) Statutes. ISTO. Brussels. Available at: www.oits-isto.org/oits/files/resources/699en.pdf (accessed 5 September 2018).

Jolin, L. (2003) Social tourism, a rich concept of its evolutions. [Le tourisme social, un concept riche de ses évolutions]. *Le Tourisme Social dans le Monde* 141(3°), 6–8.

Ley Federal de Turismo (1984) Mexico City, editorial PAC, S. A. de C. V.

Lominé, L. and Edmunds, J. (2007) *Key Concepts in Tourism, Palgrave Key Concepts*. Palgrave MacMillan, London.

Muñiz, D. (2001) *Social tourism policy. [La Política del Turismo Social]*. Universidad de Málaga. Consejería de Turismo y Deporte de la Junta de Andalucía, Seville, Spain.

Rodríguez, M. (1982) *Tourism General Theory. [Teoría General del Turismo]*. South-Western Publishing Co, Cincinnati, Ohio.

Secretaría de Turismo (Sectur) (2000) Great Vision Tourism Study in Mexico: 2020 Perspective. [Estudio de Gran Visión del Turismo en México: Perspectiva 2020]. Sectur, Mexico City. Available at: https://cedocvirtual.sectur.gob.mx/janium/Documentos/4858_002.pdf (accessed 22 October 2018).

Secretaría de Turismo (Sectur) (2006) *Tourism for All State Council. Integration and Management Report. [Consejo Estatal de un Turismo para Todos. Manual de Integración y Gestión]*. Sectur, Mexico City, p. 5.

Tonini, N. (2006) *Social Tourism: A Challenge for the New Europe*. Bureau International du Tourisme Social, Vilnius, Lithuania, p. 3.
Torruco Marqués, M. (1988) *Tourism institutional history in México 1926–1988. [Historia institucional del turismo en México 1926–1988]*. Secretaría de Prensa y Difusión, Asociación Nacional de Egresados de Turismo, Mexico.
United Nations (UN) (1948) Universal Declaration of human rights. Available at: www.un.org/en/universal-declaration-human-rights/index.html (accessed 5 September 2018).
World Tourism Organization (WTO) (1980) Manila Declaration on world tourism. UNWTO Declarations, 6 (4). Available at: www.e-unwto.org/doi/pdf/10.18111/unwtodeclarations.1980.6.4.1 (accessed 7 September 2018).

3 Challenges and Opportunities for Development of Social Tourism in Finland

Raija Komppula* and Elli Vento
University of Eastern Finland, Business School, Joensuu, Finland

3.1 Introduction

In Finland, social tourism is seen as a part of the national social policy. A social holiday is considered as a manifestation of social tourism, which is defined as representing the moral perspective of social tourism, 'Tourism for All' (Minnaert *et al.*, 2006; McCabe and Diekmann, 2015; Komppula and Ilves, 2018), which is a priority for the International Social Tourism Organisation (ISTO, 2018) and for the social tourism sector generally. According to Eurostat material deprivation statistics (Eurostat, 2018), 15% of Finnish citizens (825,000 individuals) cannot afford a week's annual holiday away from home. Social holidays are meant for people who, mainly because of financial or health-related issues, would not otherwise have a chance to go on a holiday (LYR, 2018). Social tourism is financed through profits of Veikkaus Oy (2018), which is a gaming company and a government monopoly owned by the Finnish State. The Funding Centre for Social Welfare and Health Organisations (STEA, 2018), which operates under the umbrella of the Ministry of Social Affairs and Health, is responsible for the management, allocation and supervision of the funding.

Finnish social tourism represents the adaptation model (Minnaert *et al.*, 2011), as a social holiday is intended for a certain target group and adapted to the needs of this group. It can also be characterized as visitor-related, since the activity first and foremost benefits the disadvantaged customer (Minnaert, 2007). In the Eurostat (2018) indicators, the lack of a holiday away from home is seen as an indicator of poverty, which would demonstrate that an opportunity for a holiday is acknowledged to be a part of well-being. Nevertheless, Minnaert *et al.* (2011) recall that social tourism may also provide economic opportunities for the host community by increasing sales and employment opportunities in the destination. This so-called stimulation model of social tourism or host-related social tourism that benefits the host region has a more central role in recent social tourism interpretations, offering a financial rather than purely moral argument for social tourism development (Minnaert, 2007; Minnaert *et al.*, 2011). In the Finnish context, the stimulation model and/or the host-related social tourism has not been discussed so far.

Although social tourism has a long tradition in Finland, it has been neglected in socio-political discussions. Also, the research on social tourism is still in its infancy. The major shortcoming of previous research is the descriptive level of data analysis, or cursory report of the results. The reason for this might be that most studies have been conducted to satisfy the managerial needs of the social tourism organizations, in this sense serving their purpose. The studies of Pietilä (1997, 1998) focus on the effects of

*Corresponding author: raija.komppula@uef.fi

 Social Tourism: Global Challenges and Approaches (eds J. Lima and C. Eusébio)
DOI: 10.1079/9781789241211.0004

social holiday activities, and a study by Rissanen *et al.* (2001) reports the customers' expectations towards a social holiday referring to joy, exercise, peer support and coping amid the chores of daily life. A review published in 2005 addresses the life situation of people applying for subsidized holidays, their need for respite and their wishes regarding their holiday (Kinnunen and Puhakka, 2005). In more recent academic studies, Komppula *et al.* (2016) have investigated the formation of the social holiday experience, concentrating on the role of arranged activities. Komppula and Ilves (2018) have also examined the social holiday experience formation from the holiday taker's point of view. It should be noted that the views of the social tourism operators have not been surveyed in the previous analyses.

This chapter will introduce a case study concerning the practices of social tourism in Finland, with a focus on challenges and opportunities for development. Representatives of four social tourism associations and STEA were interviewed, and their perceptions of the major challenges and opportunities for the future development investigated. Also, the interviewees' opinions of the stimulus model/host-related social tourism, as well as some alternative forms of social tourism implementation that have not been adopted in the Finnish system, were surveyed. The understanding gained can create a basis for, for instance, academic and sociopolitical discussions, further social tourism analysis, and future development and cooperation inside the operational field.

3.2 The Current System of Social Tourism in Finland

3.2.1 Brief history and the current system

The origins of Finnish social tourism lie in civic activity and date back to the 1940s, when the first social tourism organization was founded after the inception of the Annual Holidays Act in 1939 (Kinnunen and Puhakka, 2005). The first social tourism organization was accepted as a member of the Slot Machine Association (RAY), which was responsible for the allocation of funding till 2016. The current Veikkaus Oy was formed in 2017, when the previous three organizations, Fintoto, RAY and Veikkaus, were merged as a single operator. In Finland, profits from harmful gaming activity have traditionally been targeted at good purposes (e.g. health and welfare organizations, culture and arts, sports and science), which can be justified by moral aspects and societal benefit (Veikkaus Oy, 2018). Nowadays STEA, a governmental office under the umbrella of the Ministry of Social Affairs and Health, controls all Veikkaus funding that is distributed to health and welfare organizations. As a result, STEA also monitors the activities of five holiday associations. Annually, STEA (2017) allocates more than €300 million in funding for approximately 800 different welfare and health organizations, of which about €15 million is distributed to social tourism.

Little by little, the operational field of social tourism, which used to be composed of various small and often local associations, widened and evolved (Kinnunen and Puhakka, 2005). The first target groups of Finnish social tourism were mothers, who often lived in the countryside and had large families, and workers with low salaries, who did not normally have a chance to go on holiday (Kinnunen and Puhakka, 2005). Back in those days, holiday associations had connections to the political field, political parties, labour movement and agricultural organizations (Kinnunen and Puhakka, 2005). This was a very typical setting in the past society.

Until 2012, the demand for social tourism had been more or less taken for granted for social tourism associations, which could send their clients to their preferred holiday centres (Komppula *et al.*, 2016). Most often these holiday centres, which were typically holiday villages with cabins that were not suitable for year-round use, were owned by the same organizations. In 2012, a new policy was launched, which requires associations to introduce open competitive bidding for all service providers offering holiday services. In the competitive bidding, the holiday association defines the target groups for which it will offer the holidays, as well as the qualifications for the holiday programmes, according to which the service providers formulate their offer (Komppula *et al.*, 2016). Owing to the new situation, several small holiday associations with similar political backgrounds merged during

2012–2013 and formed the current three large and two small holiday associations.

3.2.2 Social holidays from the customer's point of view

In Finland, a social holiday typically lasts for 5 days and includes full board as well as a wide range of organized activities run by instructors (Komppula and Ilves, 2018; LYR, 2018). All social holidays are organized in domestic destinations and participation in activities is voluntary (Komppula and Ilves, 2018). A person who is interested in applying for a social holiday can send his/her application form to any of the holiday associations (LYR, 2018). In principle, the criteria for selecting candidates for the holidays are the same for all associations, but depending on the strategy and the main target groups of the association, some differences in priorities in terms of criteria may occur. All organizations must consider social, financial and health-related factors, when evaluating the need for the support (LYR, 2018). There is no fixed model for the evaluation process, but the associations can use different combinations of the named factors and STEA monitors how the defined criteria are reflected upon in the customer selection. Earlier participation in social holidays is also taken into account, and first-time applicants must be given priority. The holiday associations have a common electronic application and customer selection system, which enables them to assure that the same applicant cannot get support from several organizations at the same time or be accepted on a holiday too often (LYR, 2018).

The holiday associations receive approximately 45,000–50,000 applications per annum, of which approximately 30% are accepted in the customer selection process. The deductible costs of a social holiday may vary from zero to about €100 per person, depending on the holiday association and the target group (LYR, 2018; Hyvinvointilomat ry, 2018a). Customers must also organize and finance the transportation to the holiday destination by themselves. Families that cannot afford the deductible costs may apply for economic support from their home municipality or the local parish (Hyvinvointilomat ry, 2018a). The Social Welfare Act (1301/2014, 28§) (The Social Welfare Act 1301, 2014) legislates the right to have support for a holiday based on the criteria of family situation, long-term illness or disability. Nevertheless, the right for support is not a subjective right, but the decision of support for a holiday is based on the assessment of needs and discretion power of the responsible social worker, within the limits of the municipality social policy. The legislation enables municipalities to support social holidays, but the scarcity of resources and municipal guidelines may prevent support other than strictly necessary needs, such as accommodation and food.

In the Finnish context, the defined aims of social holiday activities include promoting opportunities to participate in and contribute to activities, strengthening social networks and creating a sense of community, supporting life management, improving intra-family communication, and developing life patterns conducive to health and emotional well-being (e.g. Hyvinvointilomat ry, 2018b). Social holidays are also determined as a precautionary activity, which refers to preventing problems before they occur (Hyvinvointilomat ry, 2018b). Typically, social holidays are designed for a particular target group and/or around some theme. In so-called general holidays, the target groups are commonly wide (e.g. families with children, work-aged individuals and seniors) (Hyvinvointilomat ry, 2018a; LYR, 2018). Theoretically, anyone can apply for these holidays and, for example, no special diagnoses are needed. So-called group holidays are more specified and intended for particular customer groups (e.g. individuals suffering from a certain health issue, disability or illness, unemployed people, immigrants or pensioners) (Hyvinvointilomat ry, 2018c). Group holidays are commonly executed in cooperation with different associations and organizations related to the target group of a holiday (Hyvinvointilomat ry, 2018c). These associations may be nationwide or regional.

3.2.3 Visitor-related vs. stimulation model of social tourism

In its current form, Finnish social tourism can be seen as representing visitor-related social tourism, since it benefits the holiday takers and answers to their needs (Minnaert *et al.*, 2006; Minnaert, 2007). Additionally, holidays are designed for certain target groups, which relates the concept to the adaptation model of social tourism (Minnaert *et al.*, 2011). Although the themes and target groups of social holidays vary, the concept in itself has become rather fixed inside the operational field, and the great majority of all social holidays represent the typical 5-day form. However, the discussion of the need for more flexible forms of social tourism implementation has started. It has been suggested that as the target groups of social holidays are diverse, there should also be more individualized options available for the clients (Komppula *et al.*, 2016; Komppula and Ilves, 2018).

Currently, there is a wide variety of service providers offering social holidays in Finland, ranging from small rural holiday villages to bigger spa resorts and sports institutes (see e.g. Hyvinvointilomat ry, 2018d; Maaseudun Terveys- ja Lomahuolto ry, 2018). A great number of them are located in rural areas, typically offering nature-based social holiday activities and experiences (Hyvinvointilomat ry, 2018d; Maaseudun Terveys- ja Lomahuolto ry, 2018). In the resorts offering social holiday services, social tourism creates customer flows year-round, also during the shoulder seasons. This mitigates the effect of seasonal fluctuation of demand that is a common issue inside the tourism industry (see e.g. Minnaert, 2007; Minnaert *et al.*, 2011). When financial support for social tourism has been decreasing, it has had a negative impact on the business and profitability of service providers involved in the activity – particularly those that are dependent on the social tourism demand. Although, for instance, the Calypso – Social Tourism in the EU – project is focused specifically on social tourism as a tool for strengthening the long-lasting employment opportunities and revenue generation potential in peripheral areas (Minnaert *et al.*, 2011), so far no active discussion on the stimulation model and/or host-related social tourism has emerged in Finland.

3.3 Methodology

The empirical data for this study were collected via five semi-structured interviews, in which the representatives of four holiday associations participated, as well as two representatives of the Funding Centre for Social Welfare and Health Organisations. One small holiday association was not involved in the interviews due to scheduling issues. The informants received the preliminary questions approximately 1 week before the interviews, but in practice they were allowed to talk openly, owing to which some of the interviews had narrative characteristics. Three of the interviews were held face to face and two via Skype. The length of the interviews varied from 1 h 44 min to 37 min. The overall duration of recordings was 6 h 2 min. All interviews were transcribed, and the data analysed by theme-based content analysis.

First, the interviewees were asked to talk about the major challenges and threats of social tourism from the perspective of their own organization and the Finnish social tourism system in general. Then, their views concerning positive aspects and future opportunities were ascertained. The informants were also asked to consider if it would be possible to define social tourism as a tool of regional development policy that supports the survival of rural and peripheral areas. Additionally, alternative forms of social holidays from other countries, such as holiday vouchers that allow the beneficiaries to choose their service providers and consume the services independently (Minnaert *et al.*, 2011; Puczkó and Rátz, 2011), were introduced to them. Interviewees were asked to consider if this or some other new forms of social holidays would be applicable in Finland.

3.4 Findings

In the following sections, the findings of this study are presented. The findings are organized in themes that were discussed during the interviews. The topic areas were brought forth either directly, when the interviewees were asked to share their views about them, or indirectly, when discussion around some other topic overlapped with the focus of this study. It should be noticed that whenever the holiday association representatives talk about 'the financier', they invariably refer to STEA.

3.4.1 Target groups and the general familiarity of social tourism

As has already been mentioned above, in Finland social holidays are intended for people who otherwise would not have the opportunity to have a holiday away from home. Nowadays, so-called first-timers also have the strategic priority when social holidays are allocated to the applicants. The most important separate target groups of Finnish social tourism are currently families with children, individuals and families suffering from a certain health issue, illness or disability, lonely individuals, seniors and unemployed people. Other examples of target groups are caregivers and immigrants. The representatives of the holiday associations relate that no matter what target group is in question, their customers typically have low income. Additionally, various life challenges tend to accumulate, owing to which the boundaries between the target groups are often blurry. For instance, there can be health-related and financial issues, as well as unemployment, within one family.

In recent years, the financier has influenced strategic alignments concerning social holiday target groups. In addition, the role of the first-timers has been accentuated and resources have been transferred from health-related customer groups mainly to families with children. The representatives of STEA explain that the reason for this is that in Finland, people with health issues typically have the possibility of participating in various activities and sources of support related to their condition. Some examples of these are rehabilitation financed by the social insurance institution of Finland, adjustment training, and activities organized by different associations and groups of volunteers. In other words, the financier wants to target social holidays more efficiently at those who do not have a wide range of other possibilities for participation. Generally, families with children have become the most important single customer group of Finnish social tourism, which seems to be accepted and considered as justified by the interviewees: one informant states that families with children typically have the biggest need for supported holidays. However, the ongoing development can also raise questions among those operators that have specialized in health-related target groups.

Although social tourism has a long tradition in Finland, the concept is relatively poorly known among the general public. This has been recognized as an issue that complicates catching the target groups, particularly the first-timers, who have not been on a social holiday before. One interviewee suspects that the need for social holidays might be recognized among those who work with, for instance, disabled people, caregivers or customers of child welfare, which may mean that these groups receive information about social tourism more efficiently than those who do not have similar authority contacts. Another holiday association representative ponders that actually the current strategic alignments are good, since the priority of the first-timers 'forced [us] to invest in marketing ... as this is not known, we must make it familiar'. One major question is whether those people who most desperately need a social holiday ever end up as social holiday customers. Sometimes a Finnish social holiday customer has been characterized as someone who somehow perseveres and hangs on in society. Although social tourism is described as low-threshold activity, for someone who is completely excluded, 'this can though be a high threshold'. The informant notes that if the potential applicant is particularly excluded, finding 'the path to social tourism' would require active cooperation between different actors.

3.4.2 Cooperation inside the operational field of social tourism

When cooperation is discussed in the context of Finnish social tourism, there are clear contradictions in the empirical data – even in the views of the individual interviewees. On one hand, cooperation between the holiday associations is described as being at a good level: 'From the year 2012, when this association structure was formed, we have had good cooperation ever since ... we negotiate and have discussions about all important matters'. The same informant states that 'In some sense our smallness ... unites us'. The representatives of STEA also note that the holiday associations cooperate actively and closely with one another. Nowadays besides the common electronic customer selection and application system, they also have the official co-operation group of holiday associations (LYR), which comes together regularly. It is also pointed out that the holiday associations seem to have good relationships and good communication 'to the field', meaning the service providers with which they organize the social holidays.

On the other hand, the operational field of social tourism is described as 'siloed', referring to a situation in which operators 'stay in their own siloes' without properly communicating and sharing ideas with one another. One interviewee ponders that to some extent the holiday associations may be protecting their own services and image, whereby sharing these aspects with others can feel challenging. Another informant explains that sometimes there has, for example, been a situation in which customers of different holiday associations have been on a holiday at the same time in the same place, and problems have occurred, when one association has offered its customers more services or different services from the other one: 'and that is a terrible situation, when the group is divided like "you get this [service] and you do not"'. The interviewee emphasizes that this kind of situation should be avoided by communication and cooperation.

As well as the silo effect existing between holiday associations, it is claimed that it is also in effect between other operators that the associations should work with to be able to catch the right target groups: 'As a holiday association we do not reach those who would need [a social holiday] the most; to be able to do that we kind of need agents and cooperatives'. It is noted that this cooperation should be developed towards a partnership, in which aims and goals are genuinely shared between the actors. The interviewee contemplates that although this kind of a setting is challenging to achieve, 'there is also a lot of potential' that has not yet been harnessed.

3.4.3 Perceptions of control and the question of future funding

Generally, notable differences in perspectives, particularly referring to the amount of funding, reporting and control, can be recognized between the holiday association representatives and the representatives of STEA. The holiday association representatives state that in recent years, the funding distributed to social tourism has been decreasing. All interviewees point out that in Finland, poverty and societal inequality has deepened and will deepen in the future, owing to which 'the rich become richer and the poor become poorer', and there will be a growing need for social holidays, particularly for families. One informant thinks that whenever the financier is distributing less money for health and welfare associations, social tourism seems to be the first entity whose funding is cut.

Besides decreasing funding, holiday association representatives also criticize the financier's increased control and supervision, as well as the bureaucracy connected to it. One interviewee describes how 'this application, accounting and other reporting ... like this paperwork, bureaucracy, well ... this feels exorbitant. And what is even more exorbitant, it feels like it is just increasing here.' The informants state that to some extent the required reporting is twofold, and one respondent even ponders that currently there might be trust issues between the financier and the holiday associations: 'perhaps they do not trust that we are doing this work properly'. The interviewee also notes that 'the trust that we had from RAY [the former organization responsible for funding and monitoring], at least I feel like it has gone somewhere else'.

The representatives of STEA, on the other hand, emphasize that they have not been completely happy with the way that the holiday associations have reported the effects of their actions so far. Generally, STEA is interested in the need, necessity, effectiveness, variety and quality of the activities that they are funding. The informants point out that, so far, reporting on effects by holiday associations has often been conducted superficially and in a descriptive way, although, according to the results of some previous academic surveys, social holidays can have deeper-level impacts on the lives of the participants. It is put forward that if the deeper-level effects are not reported, it may mean that they do not exist, owing to which the need for funding is not being justified and the resources distributed to social tourism could perhaps be 'used for something else'. This is an important statement, since to avoid addictions and other harmful impacts of gaming, some new procedures have been planned for the Veikkaus gaming system. It has been forecast that these new procedures would decrease the profits of Veikkaus Oy at least momentarily and intensify the competition for the remaining funding resources.

Although the representatives of the holiday associations criticize the current level of control and bureaucracy, particularly in the situation in which funding has been decreasing, there has recently been some activity to better respond to the need for more efficient reporting of effects of the holidays, either in the form of official research projects or as customer surveys designed by the associations. One interviewee describes how the results of their recently developed customer surveys have so far been 'surprising' and 'better than expected'. For example, more than 50% of the respondents reported that 6 months after a social holiday they exercise either a lot more or somewhat more than before the holiday, and 90% have reported that their family connection has improved. The informant ponders that social tourism may have 'great societal impacts ... which we currently cannot measure directly, but they exist anyway'. The informant also notes that social tourism 'can never eliminate exclusion, but it can have a lot bigger role than the actual scope of the activity in decreasing exclusion and promoting the well-being of people'.

3.4.4 Stimulation model/host-related social tourism

When the informants discuss the possibility of using social tourism as a regional and/or industrial policy tool that supports the well-being of peripheral and rural areas, as well as compensating for the seasonal fluctuation in tourism demand, many of them note that to some extent this already occurs due to the open competitive bidding – although the development has not been intentionally guided. In terms of competitive bidding, 'Those holiday resorts that are interested in this activity participate. And it is that certain places that have a certain image do not participate in this competitive bidding.' It can be seen that, for example, certain spa resorts want to be profiled as high-quality places, and as such they do not want to have social tourism customers. Additionally, the big ski resorts in Lapland do not participate in the social tourism services competitive bidding, since they have enough self-paying customers anyway. If a holiday resort that is not very popular among the customers becomes selected in the competitive bidding, a holiday association may have difficulties in having enough applications that can be accepted in the selection process. Sometimes holidays are even cancelled owing to a lack of applicants, which does not benefit anyone. However, it is also pointed out that the Finnish tourism sector is currently in a difficult situation, owing to which holiday associations have 'surprisingly high-quality and fine places for a low price'.

The idea of intentionally targeting social tourism customer flows to peripheral regions also generates some criticism. One respondent explains that this would have a negative effect and would differentiate disadvantaged individuals from ordinary self-paying customers: 'Like "well, let's fill [the resorts] with those people then".' The representatives of the governmental body point out that STEA does not have any industrial political perspective in their actions. Additionally, they note that if the holiday resorts are located in peripheral areas, 'No one gets there. Then no one wants to go on a holiday and then the funds are wasted when they cannot be used ... in this sense it is of course in the interest of STEA that the holiday resort that wins in the competitive bidding is the kind where people want to go, so that the funds are utilised.' Generally, it has been noticed that inaccessibility can be an issue if there is no public transport to the destination.

In terms of scheduling, social holidays often take place during weekdays, which are 'worse for those hotels', since at weekends the demand is typically higher. It is also pointed out that paying weekend accommodation prices is not profitable, because the holiday association would get fewer holidays in return for their money. The informants state that family holidays must be organized during school holiday periods, which are typically in high season in the tourism industry. However, for example, for some pensioner groups and families with small children, the date of a holiday is not necessarily a significant matter, and those holidays are organized year-round. The representatives of STEA emphasize that they do not guide the scheduling of social holidays, but that holiday associations can organize them whenever they want to.

3.4.5 Potential new forms of social tourism implementation

Although the governmental body in Finland does not restrict the duration of social holidays, the great majority of them represent the traditional 5-day format. Some holiday associations have also organized shorter holidays or holidays split into two separate periods, which can be suitable, for instance, for caregivers and families with pets that cannot be left alone for several consecutive days. However, one representative of a holiday association believes that their customers would probably have difficulty covering the transportation costs of two trips if a holiday took place in two periods. The interviewee also believes that a longer holiday can have a more positive impact in terms of recreation. Particularly for those customer groups that have special needs, there have sometimes been voluntary support/peer support people recruited at the holiday destination. These support people spend time with and help the holiday makers during the holiday. Some associations also organize common holidays for caregivers and their patients. At the holiday destination, there are professional instructors on the spot who can look after the patients so that the caregivers can have a break and spend time by themselves.

When the interviewees are asked how social holidays could be developed in the future, many of them note that there could be more, shorter holidays and holidays of different durations for different target groups. Additionally, closer cooperation with different health and welfare organizations and operators related to the customer groups is brought to the fore. One holiday association representative shares an example of their recent cooperation with a group of immigrants: 'When the holiday is in midsummer, we organise a meeting there [at the holiday resort] in spring. Like the leader of the group, it is important for them that the leader comes to the destination and there we check the facilities, go through some things and we take account their special needs that can be related to food, programme, schedules...' According to some views, the Finnish social holiday system is already well developed and does not require any major reforms.

One interviewee suggests that there could be something developed for those accepted social holiday customers who are awaiting a holiday. For instance, if someone is suffering from a certain medical condition and is not a member of the patient association, the person could be informed about these participation possibilities. Additionally, the interviewee ponders that there could also be an association-based supporting procedure for those situations, in which actors become worried about a particular customer during the holiday: 'like what [support] could be offered to him/her after the holiday'. The informant clarifies the idea by explaining that the entrepreneurs and service providers are not interested in what happens to the customers after they leave the holiday resort, but the responsibility of executing the procedure should be association-based: 'In this sense, a kind of ethicalness is built into this [activity].'

Voucher-based social holidays

When voucher-based social holidays are discussed, the representatives of STEA underline that, although they support innovativeness per se, this kind of activity would not fit the current legislation and taxation system. Also, the holiday association representatives criticize the idea. First, it is pointed out that the Finnish social tourism system differs from ordinary tourism, since, for example, it involves elements related to teamwork, peer support and being in a group. Additionally, the effects of social holidays must be measured, and this would not be possible if the customers went on a holiday independently. One informant notes that voucher-based holidays could be organized via municipalities, and the holiday associations would not be needed in this process. Another ponders that 'if this would work somewhere, it would be on family holidays', since family holiday customers typically have the lowest participation rates in the organized activities. However, the same interviewee points out that there would still be a lack of peer support and supervision, which are sometimes needed during social holidays.

3.5 Conclusions

The Finnish social tourism system has recently gone through some major reforms. For a long time, actors were allowed to operate quite freely and as a result adapting to the new situation requires flexibility and development from all parties involved in the process. The current setting inside the field of Finnish social tourism involves some characteristics of the theory of organizational change, and indicates the challenges of going through and adapting to the externally driven change that requires the ability for short-run adaptation (see e.g. Weick and Quinn, 1999; Piderit, 2000; Bovey and Hede, 2001). To some extent, change resistance and a certain willingness to maintain and protect the operational field and the traditional procedures (Bovey and Hede, 2001; Hislop, 2009) of social tourism can be recognized in the findings. For example, it is noted that the existing silo effect creates challenges in terms of cooperation and development, but simultaneously, the unharnessed development potential and possibilities are also recognized. One informant states that the new strategic alignments have actually forced the actors to make positive changes and develop their work, which demonstrates the widening of the horizons and the idea of organizational adaptation (Weick and Quinn, 1999; Piderit, 2000; Bovey and Hede, 2001).

The question concerning the future funding of social tourism is urgent and critical and emphasizes first and foremost the ability to justify the need for social tourism with more efficient effect reporting. Many of the interviewees note that social tourism can hypothetically have deeper-level positive impacts on the life of the holiday makers – these impacts are not measured effectively in the current situation. This accentuates the need for social tourism research and brings forth a practical example of the social tourism-related research gap that has been recognized in previous studies (e.g. Minnaert *et al.*, 2011; McCabe *et al.*, 2012; Yeoman and Schänzel, 2012).

When potential new forms of social tourism implementation are discussed, the ideas of the holiday association representatives mainly consider holidays of different durations, and holidays executed and designed in closer cooperation with other social sector associations and actors to better respond to the needs of the target groups. Generally, in Finland, the social tourism system differs from those countries in which social tourism is based on the work of charity organizations (see e.g. McCabe, 2009; Minnaert *et al.*, 2011; McCabe and Johnson, 2013). The legislation, the specific aims of social tourism and other norms must be taken into account in the implementation of social tourism, which means that, for example, voucher-based social holidays could not be adapted to the Finnish context in the current situation. Regardless, it can be claimed that the views of the holiday association representatives are more limited to the traditional form of a social holiday than the suggestions presented by STEA. The representatives of the governmental body note that the work of the social holiday operators could cover the pre-, during- and post-social holiday stages, which refers to all temporal aspects of a tourist experience when reflected upon in scientific theory (Komppula and Gartner, 2013).

It can be claimed that Finnish social tourism indisputably represents visitor-related social tourism based on the moral value. However, mainly as a result of open competitive bidding, some characteristics of the host-related social tourism and the stimulation model have emerged in the Finnish system (Minnaert *et al.*, 2006, 2011; Minnaert, 2007). From the social tourism operator's point of view, it can cause certain challenges if the popularity and accessibility of a social holiday resort are not at a high level. Previously, the stimulation model of social tourism has been adopted, for instance, in some coastal regions of Spain, which are generally more densely populated and more dependent on the tourism industry than rural areas in Finland are, and thus may not have the same inaccessibility and unpopularity issues (Minnaert *et al.*, 2011; McCabe and Diekmann, 2015). However, it should also be noted that sometimes, for example, the whole of eastern Finland has been determined as peripheral, although there are some urban areas, traditional tourist destinations and bigger (social) tourism resorts that can be characterized as accessible in terms of transport connections. This indicates that in regional political discussions concerning social tourism, the perspective chosen as well as the differences between areas should be taken into account.

Acknowledgements

The authors express their gratitude to professor Riitta Vornanen, University of Eastern Finland, for her indispensable help during our journey to the world of social policy and social work practices in our social tourism research project.

References

Bovey, W.H. and Hede, A. (2001) Resistance to organizational change: the role of cognitive and affective processes. *Leadership & Organization Development Journal* 22(8), 372–382. DOI: 10.1108/01437730110410099.

Eurostat (2018) File: Population unable to afford to go for a week's annual holiday, away from home, 2014–17 (early data) – % of population.png. Available at: https://ec.europa.eu/eurostat/statistics-explained/index.php?title=File:Population_unable_to_afford_to_go_for_a_week%27s_annual_holiday,_away_from_home,_2014-17_(early_data)_-_%25_of_population.png&direction=prev&oldid=390582 (accessed 7 December 2018).

Hislop, D. (2009) *Knowledge Management in Organizations*, 2nd edn. Oxford University Press, Oxford, UK.

Hyvinvointilomat ry (2018a) Hae lomalle. [Apply for a holiday]. Available at: http://hyvinvointilomat.fi/hae-lomalle-2/ (accessed 7 December 2018).

Hyvinvointilomat ry (2018b) *Toimintakertomus 2017. [Annual report 2017.]*. Hyvinvointilomat ry, Helsinki, Finland.

Hyvinvointilomat ry (2018c) Tuetut lomat ryhmille. [Social holidays for groups.]. Available at: http://hyvinvointilomat.fi/tuetut-lomat-ryhmille/ (accessed 6 December 2018).

Hyvinvointilomat ry (2018d) Lomakohteet. [Holiday destimations.]. Available at: http://hyvinvointilomat.fi/lomakohteet/ (accessed 6 December 2018).

ISTO (2018) (International Social Tourism Organization) history. Available at: www.oits-isto.org/oits/public/section.jsf?id=1179 (accessed 6 December 2018).

Kinnunen, K. and Puhakka, S. (2005) *Kun ei ole varaa valita. Tuetulta lomalta toivoa arkeen, [When one can't afford to choose. Hope for daily life through a social holiday.]*, 1st edn. Cosmoprint, Solaris-Lomat ry, Helsinki, Finland.

Komppula, R. and Gartner, W.C. (2013) Hunting as a travel experience: an auto-ethnographic study of hunting tourism in Finland and the USA. *Tourism Management* 35, 168–180. DOI: 10.1016/j.tourman.2012.06.014.

Komppula, R. and Ilves, R. (2018) Social tourism as correlates of QOL: the case of disadvantaged people. In: Uysal, M., Sirgy, M.J. and Kruger, S. (eds) *Managing Quality of Life in Tourism and Hospitality*. CAB International, Wallingford, UK, pp. 54–69.

Komppula, R., Ilves, R. and Airey, D. (2016) Social holidays as a tourist experience in Finland. *Tourism Management* 52, 521–532. DOI: 10.1016/j.tourman.2015.07.016.

LYR (2018) Yleistä tuetuista lomista. [General information about social holidays.]. Available at: https://www.lomajarjestot.fi/yleista-tuetuista-lomista/ (accessed 6 December 2018).

Maaseudun Terveys- ja Lomahuolto ry (2018) Lomakohteet. [Holiday destinations.]. Available at: www.mtlh.fi/fi/Lomakohteet/ (accessed 7 December 2018).

McCabe, S. (2009) Who needs a holiday? Evaluating social tourism. *Annals of Tourism Research* 36(4), 667–688. DOI: 10.1016/j.annals.2009.06.005.

McCabe, S. and Diekmann, A. (2015) The rights to tourism: reflections on social tourism and human rights. *Tourism Recreation Research* 40(2), 194–204. DOI: 10.1080/02508281.2015.1049022.

McCabe, S. and Johnson, S. (2013) The Happiness factor in tourism: subjective well-being and social tourism. *Annals of Tourism Research* 41, 42–65. DOI: 10.1016/j.annals.2012.12.001.

McCabe, S., Minnaert, L. and Diekmann, A. (2012) Introduction. In: McCabe, S., Minnaert, L. and Diekmann, A. (eds) *Social Tourism in Europe. Theory and Practice*. Channel View Publications, Bristol, UK, Buffalo, New York, pp. 1–13.

Minnaert, L. (2007) Social tourism: a potential policy to reduce social exclusion? The effects of visitor-related social tourism for low-income groups on personal and family development. PhD thesis. University of Westminster, London, UK.
Minnaert, L., Maitland, R. and Miller, G. (2006) Social tourism and its ethical foundations. *Tourism Culture & Communication* 7(1), 7–17. DOI: 10.3727/109830406778493533.
Minnaert, L., Maitland, R. and Miller, G. (2011) What is social tourism? *Current Issues in Tourism* 14(5), 403–415. DOI: 10.1080/13683500.2011.568051.
Piderit, S.K. (2000) Rethinking resistance and recognizing ambivalence: a multidimensional view of attitudes toward an organizational change. *Academy of Management Review* 25(4), 783–794. DOI: 10.5465/amr.2000.3707722.
Pietilä, M. (1997) *Tutkimus sosiaalisesta lomatoiminnasta. [A study on social tourism.]*. LYR/Edita, Helsinki, Finland.
Pietilä, M. (1998) *Sosiaalisen lomatoiminnan jatkotutkimus: Erikoisryhmätoiminta. [Further study on social tourism: activities with special groups.]*. LYR/Edita, Helsinki, Finland.
Puczkó, L. and Rátz, T. (2011) Social tourism in Hungary: from trade unions to cinema tickets. *Current Issues in Tourism* 14(5), 459–473. DOI: 10.1080/13683500.2011.568055.
Rissanen, P., Karjalainen, S. and Salmela, T. (2001) *Kenellä on oikeus lomaan? Avustetun lomatoiminnan tuloksia. [Who has a right to a holiday? Results of subsidized tourism.]*. Sosiaalikehitys Oy, Domus-Offset Oy, Tampere, Finland.
STEA (2017) Avustustoiminnan linjaukset vuosille 2017–2019. [Funding guidelines for the years 2017-2019.]. Available at: https://www.stea.fi/documents/2184241/2348706/Linjaukset/51fab9b5-6c29-4554-ba5a-c0939bf8cbd2 (accessed 6 December 2018).
STEA (2018) Funding centre for social welfare and health organizations. Available at: www.stea.fi/web/en/frontpage (accessed 6 December 2018).
The Social Welfare Act 1301 (2014). Available at: www.ilo.org/dyn/natlex/natlex4.detail?p_lang=en&p_isn=101111 (accessed 7 December 2018).
Veikkaus Oy (2018) Avustuskohteet. [Targets of support.]. Available at: https://www.veikkaus.fi/fi/yritys#!/avustukset/avustuskohteet (accessed 7 December 2018).
Weick, K.E. and Quinn, R.E. (1999) Organizational change and development. *Annual Review of Psychology* 50(1), 361–386. DOI: 10.1146/annurev.psych.50.1.361.
Yeoman, I. and Schänzel, H. (2012) The future of family tourism: a cognitive mapping approach. In: Schänzel, H., Yeoman, I. and Backer, E. (eds) *Family Tourism: Multidisciplinary Perspectives*. Channel View Publications, Bristol, UK, pp. 171–193.

4 Benefits of Social Tourism Programmes for Seniors: The Case of the INATEL Foundation in Portugal

Maria João Carneiro[1]*, Celeste Eusébio[1], Elisabeth Kastenholz[1] and Helena Alvelos[2]

[1]Department of Economics, Management, Industrial Engineering and Tourism, University of Aveiro and GOVCOPP Research Unit, Aveiro, Portugal; [2]Department of Economics Management, Industrial Engineering and Tourism, University of Aveiro and Centre for Research and Development in Mathematics and Applications, Aveiro, University of Aveiro, Portugal

4.1 Introduction

Seniors nowadays represent an undeniably significant market in tourism that is showing considerable growth. Considering Europe, which is the world region that generates the most international tourism arrivals, and taking data from EUROSTAT (2019a) into account, it can be observed that, in the EU-28 countries, the people aged 65 or older increased by around 13,292,432 between 2010 and 2018 (an increase of 15%), representing now about a fifth of the corresponding population. In 2014, around 20% of the bed nights spent by tourists living in the EU in the context of travel for private purposes are attributable to individuals aged 65 years or more, while 39% corresponded to overnights for tourists aged 55 years or older (EUROSTAT, 2019b).

Although leisure and holidays are considered to be human rights in the Universal Declaration of Human Rights (United Nations, 1948), and the Global Code of Ethics for Tourism states that everyone has the right to tourism (UNWTO, 2019), many seniors are still confronted with several barriers when considering travelling (Fleischer and Pizam, 2002; Huang and Tsai, 2003; Kazeminia *et al.*, 2015; Huber *et al.*, 2018; EUROSTAT, 2019b). Therefore, social tourism plays a major role in enabling seniors' participation in tourism. The INATEL Foundation in Portugal and the Institute for Older Persons and Social Services (IMSERSO) in Spain are examples of organizations that design and manage social tourism programmes for seniors that may bring very important benefits to this market segment. Nevertheless, there is scant research on the benefits obtained from social tourism programmes for seniors. Moreover, some of the studies undertaken examine the impacts of these initiatives on tourism destinations (Eusébio *et al.*, 2016; Cisneros-Martínez *et al.*, 2018) and only a few analyse benefits for the seniors participating in the programmes (Morgan *et al.*, 2015; Medarić *et al.*, 2016; Sedgley *et al.*, 2018).

*Corresponding author: mjcarneiro@ua.pt

DOI: 10.1079/9781789241211.0005

The present chapter aims to extend previous research by analysing the social programmes designed for seniors offered by the INATEL Foundation over the years, and by examining the benefits of participating in these programmes, as perceived by seniors. To accomplish these aims, the chapter is structured into four sections. After this introduction, a literature review concerning the relevance of social tourism programmes in overcoming some travel constraints for seniors and the potential benefits obtained by seniors from participating in these programmes ispresented. In the third section, the social tourism programmes offered by the INATEL Foundation are described in terms of their most important characteristics, demand for them and the benefits obtained from them by the participants (seniors). Finally, the chapter ends with the most important conclusions regarding contributions to developing social tourism initiatives to increase the participation of seniors in tourism activities and further research on this topic.

4.2 Literature Review

4.2.1 The relevance of social tourism programmes for seniors in overcoming travel constraints

One of the major barriers to travel for seniors is financial constraints (Fleischer and Pizam, 2002; Alén *et al.*, 2017; EUROSTAT, 2019b), since many seniors cannot afford leisure travels. Hence, the level of income shows a significant positive influence on seniors' propensity to take a holiday and on the number of holiday days taken (Fleischer and Pizam, 2002). In 2014, 37% of people of 65 years or more stated that they did not travel owing to financial reasons (EUROSTAT, 2019b). Social tourism programmes for seniors, funded by governmental entities or other kinds of organization that offer trips at a lower price, either fixed or differing according to criteria such as the monthly average income (IMSERSO, 2019a, b; INATEL, 2019c), are the only opportunity to undertake tourism trips for many seniors.

For some seniors, the lack of someone to travel with is also a major constraint on participating in leisure trips (Huang and Tsai, 2003; Huber *et al.*, 2018). This difficulty may derive from these seniors' limited social networks as well as from other situations, such as having lost their partner (Huber *et al.*, 2018). Therefore, trips in social tourism programmes are of major relevance for these seniors as they give them the chance to travel in a group of other seniors with whom they can interact.

The fear of not being able to deal with all the requirements of a trip, both at the planning stage, while choosing or booking transportation and accommodation, and during the trip, when confronted with the need to find the way to places or to talk with foreigners, among other features, also prevents some seniors from travelling (Huang and Tsai, 2003; Kazeminia *et al.*, 2015). Engaging in social tourism programmes minimizes planning requirements since the organizations that manage the programmes carry out booking and other planning activities. Furthermore, the presence of someone representing these organisations during the trip and accompanying participants (INATEL, 2019b, c) decreases seniors' level of anxiety and makes them feel safer.

Health problems also limit seniors' possibilities of travelling and increase their stress concerning the trip (Fleischer and Pizam, 2002; Huber *et al.*, 2018). Some seniors try to adopt coping strategies to overcome or minimize health problems that may occur during the trip; for example, taking out insurance or travelling to destinations with good medical facilities and assistance (Huang and Tsai, 2003). Most trips in social tourism programmes for seniors include insurance (INATEL, 2019b) and in some instances even include medical assistance in the hotel (IMSERSO, 2019a). Some programmes also offer menus adapted to the health needs of seniors (INATEL, 2019b). It is also of major importance to train guides or entertainers accompanying seniors to address complex health and medical issues (Daniels *et al.*, 2019).

Apart from the fact that social tourism programmes help seniors overcome substantial travel constraints, they provide this segment a wide range of benefits through participation. The next section will present a discussion on these benefits.

4.2.2 The benefits of social tourism programme trips for seniors

Dann (2002) advocates that leisure travel may increase seniors' quality of life. Hwang and Lee (2019) provide empirical evidence in this context, since in their study with senior tourists in Korea, the four dimensions of the tourism experience suggested by Pine and Gilmore (1998) – education, entertainment, aesthetics and escapism – all had a positive influence on seniors' perceived well-being. However, it is important to understand what the main benefits of tourism may be and specifically of social tourism programmes for seniors.

Analysing seniors' motivations for undertaking tourism trips may provide important perspectives on the potential benefits that seniors are looking for in tourism trips and that they may obtain from them. Although the senior market is recognized as being somewhat heterogeneous in terms of tourism motivations (Lee and Tideswell, 2005; Shavanddasht, 2018), some motivations seem to have a major role in seniors' decisions to travel. In a review of 36 studies on travel motivations of senior tourists from 1980 to 2017 carried out by Otoo and Kim (2020), the five most important motivation domains for senior tourists are, in descending order: (i) destination appeal; (ii) socializing; (iii) knowledge acquisition; (iv) novelty and escape; and (v) rest and comfort. A more recent work (Pestana *et al.*, 2020) and a qualitative study not included in the review (Small, 2003) also corroborate some of these findings.

Looking in further detail at the motivations previously identified, resting and relaxing seem to be important travel motivations for seniors (Otoo and Kim, 2020), although they sometimes seem to assume higher relevance among younger seniors (Pestana *et al.*, 2020), probably because they are more physically active at home and sometimes in their jobs. Tourism may provide relaxation benefits for seniors, for example by permitting closer contact with nature, engaging in pilgrimages that can help obtain peace (Pezeshki *et al.*, 2019) or travelling to places with a relaxing atmosphere. Various trips in social tourism programmes for seniors that have rural or nature destinations where people can appreciate natural landscapes, or thermal spring destinations with a calm environment (IMSERSO, 2019a, b; INATEL, 2019c) may help provide the benefits of relaxation and rest.

Furthermore, the possibility of travelling to a different place and meeting new people offered by these programmes, as well as the positive atmosphere created during the trips (Sedgley *et al.*, 2018), is revealed as an excellent means for seniors to escape from the boredom of their daily routines (Moal-Ulvoas and Taylor, 2014; Morgan *et al.*, 2015; Eusébio *et al.*, 2017; Sedgley *et al.*, 2018), from stress, heavy responsibilities and negative psychological states experienced in day-to-day life, thereby restoring their psychological balance (Morgan *et al.*, 2015; Sedgley *et al.*, 2018).

Seniors also appreciate tourism as an opportunity to have new experiences and get to know new places (Eusébio *et al.*, 2017; Pestana *et al.*, 2020), and this group shows great interest in visiting historic and cultural attractions and learning about different cultures and lifestyles (Otoo and Kim, 2020; Pestana *et al.*, 2020), potentially even increasing their pride in their own country of residence (Pezeshki *et al.*, 2019). Social tourism programmes offer seniors the opportunity to participate in trips to a wide range of destinations (IMSERSO, 2019a; INATEL, 2019c), generally more than some seniors could have access to by themselves, with some destinations being especially culturally appealing. In addition, some of these initiatives also include cultural entertainment activities (INATEL, 2019b) that promote contact with typical customs and traditions of the destination visited.

Socialization, including developing friendships, emerges as one of the main motivations for seniors to participate in tourism trips (Small, 2003; Carneiro *et al.*, 2013; Eusébio *et al.*, 2017; Otoo and Kim, 2020; Pestana *et al.*, 2020). This is probably also a consequence of the limited social networks and isolation that are a hallmark of the life of many seniors (Huber *et al.*, 2018). Travelling in a group within social programmes provides good opportunities for interaction with travel companions, while additionally these programmes often provide opportunities for socialization at several moments on the trip (Medarić *et al.*, 2016; INATEL, 2019b) and specifically during entertainment activities, where the chance to interact with local residents is sometimes offered. In addition, some programmes let participants take another person of their choice on their trip (Medarić *et al.*, 2016; INATEL, 2019b), permitting seniors to strengthen bonds with a relative or friend who accompanies them. These programmes are thus a powerful instrument for reinforcing and extending senior citizens' social networks.

Given these opportunities for social contact, escape from routines and getting to know new places, these programmes sometimes also have a major role in broadening seniors' horizons (Moal-Ulvoas and Taylor, 2014). Some seniors reveal that they have time to reflect about themselves during travel, and that visiting other countries with worse living conditions and interacting with others gave them a more positive perspective about their own life and a better understanding of other people (Moal-Ulvoas and Taylor, 2014).

4.3 Social Tourism Programmes for Senior Citizens Organized by the INATEL Foundation

4.3.1 The role of the INATEL Foundation in the development of social tourism

The development of social tourism in Portugal began with the creation of the National Foundation for Joy and Work [*Fundação Nacional para a Alegria do Trabalho* – FNAT], in 1935, with the goal of developing leisure activities for employees. However, it was only after 1948 that social tourism underwent an important development with the Universal Declaration of Human Rights recognizing that 'everyone has the right to rest and leisure, including reasonable limitation of working hours and periodic holidays with pay' (United Nations, 1948, art. 24).

In 1956, FNAT participated in the first international congress about social tourism in Bern, where the concept of social tourism was institutionalized. Owing to differences regarding the development of social tourism from country to country, the Bureau International du Tourisme Social (BITS) was created in 1963, including public organizations as well as independent private organizations with activities in the field of social tourism (Minnaert *et al.*, 2012). In 1971, FNAT became a member of BITS.

In 1975, FNAT came to be called the National Institute for Workers' Leisure Time Occupation [*Instituto Nacional para o Aproveitamento dos Tempos Livres dos Trabalhadores*] (INATEL), acting in the provision of social services, with an emphasis on culture, sports and social tourism. From then on, the tourism activities organized by INATEL addressed all segments, with a special focus on seniors, through health and well-being tourism programmes.

In 1995, the Portuguese government promoted the development of social tourism programmes for seniors, conceding the management of these programmes to INATEL. Since this date, various social tourism programmes for seniors have been developed and a great number of Portuguese citizens have benefited from these programmes, as can be seen in Section 4.3.2 of this chapter.

The social tourism programmes undertaken aim to give senior citizens the opportunity to participate in tourism activities and occupy their spare time, at reduced prices and with excellent safety and assistance conditions (Eusébio *et al.*, 2016). Moreover, these programmes also have a crucial role in enriching the participants' culture, improving their quality of life and well-being, and avoiding isolation and exclusion (Eusébio *et al.*, 2013). However, these programmes also have an important role in the economic development of Portuguese tourism destinations through both direct and multiplier effects (Eusébio *et al.*, 2013, 2016).

In 2008, INATEL ceased to be part of the central administration with the creation of a private–public benefit foundation – the INATEL Foundation (Decree law no. 106/2008). The main aim of this organization is to promote the best conditions for the occupation of free time and leisure for active and retired workers, developing social tourism initiatives and cultural, recreation and sporting activities in order to promote solidarity and social inclusion.

The INATEL Foundation is a member of the International Social Tourism Organisation (ISTO), and has promoted, administrated, managed and financially supported various social tourism programmes directed to the four target markets of social tourism initiatives (young people, seniors, people with disabilities and families with low income or difficulties) (UNWTO, 2020). However, the financial support is not directly transferred to the tourist but is provided through discounts on the tourism trips offered.

Despite the important role of the INATEL Foundation in the development of social tourism in Portugal for all of its target groups, the elderly population is the target examined in this chapter. Therefore, in the next section only INATEL social tourism programmes for seniors will be characterized.

4.3.2 Characterization of the senior tourism programmes developed by the INATEL Foundation

Portugal has a long tradition in the offer of social tourism programmes for the senior market. Since 1995 the INATEL Foundation has been the institution responsible for the organization and promotion of social tourism programmes for this market, permitting access to tourism for low-income senior citizens, thereby reducing their isolation (INATEL, 2013). These offers started with two different programmes: (i) the Senior Tourism Programme (SETP) [Programa Turismo Sénior]; and (ii) the Health and Thermal Treatments Senior Tourism Programme (HTSTP) [*Programa Saúde e Termalismo Sénior*]. The former, initiated in 1995, was a joint initiative of the Ministry of Economy and Innovation and the Ministry of Labour and Social Solidarity. The latter, initiated only in 1997, was a joint initiative of the Ministry of Health and the Ministry of Labour and Social Solidarity.

The target of both programmes was Portuguese citizens aged 60 years or older, and the price of the trips varied according to seniors' income levels. The seniors accordingly had access to a package tour (including, for example, accommodation, transport, and recreational and cultural activities), with a corresponding discount. The trips included in the SETP programme had a duration of 8 days (7 nights), to destinations located mainly in Portugal, including accommodation, transportation, cultural and entertainment activities, and the support of a recreational assistant. However, the HTSTP additionally included thermal treatments and the trips had a duration of 15 days (14 nights). Through these two social tourism programmes, the seniors had the opportunity to socialize, improve their health and increase their knowledge.

The majority of tourism trips organized in the scope of social tourism programmes occurred in Portugal. However, since 1999, through a transnational programme of bilateral exchange between INATEL and IMSERSO, reciprocal exchange programmes between Portugal and Spain took place. Therefore, a group of Portuguese participants carried out tourism trips in Spain and a group of Spaniards carried out tourism trips in Portugal (about 4000) (European Commission (EC), 2010).

In 2008, the Programme 60+ Azores (60+A) [*Programa 60+Açores*] was launched in order to enable the participation of low-income residents aged 60 or over in the Azores Archipelago in tourism trips to the nine islands of the archipelago. This programme was an initiative of the Regional Secretariat for the Economy, Labour and Social Solidarity of the Regional Government of the Azores, and was organized and managed by the INATEL Foundation. With a duration of 8 days (7 nights), the tourism trips included flight, transfers, accommodation, cultural and tourism tours, insurance against personal accidents and continuous monitoring by a recreational assistant (INATEL, 2011).

During 2011 and 2012, the Programme 'Always on Holiday' (PAH) [*Programa Sempre em Férias*] was promoted and organized by the INATEL Foundation to stimulate the active participation of Portuguese citizens aged 60 or over in long-duration tourism trips (INATEL, 2011).

Between 2013 and 2017, a new social tourism programme – the 'Quiet Days Programme' (QDP) [*Programa Dias Tranquilos*] – directed to citizens aged 55 or over who wish to benefit from holidays and rest in hotel units of the INATEL Foundation over 15 days – was offered by the INATEL Foundation.

In 2018, a new social tourism programme for seniors – Programme INATEL 55+.pt (55+.pt) [*Progama INATEL 55+.pt*] was implemented, replacing the social tourism programme for seniors mentioned before. This new programme belongs to a new generation of senior tourism and social travel measures, with the support of European and national funds. The trips organized in the scope of this STP take place in the low season, have a duration of 6 days (5 nights) and are directed at Portuguese citizens aged

55 or over. With a rationale of equal opportunities and leisure promotion, through this STP the INATEL Foundation expects to encourage healthy living and active ageing of seniors who are at risk of social exclusion or who have low income. Therefore, the aims of this STP are to: (i) promote leisure; (ii) promote active ageing; (iii) combat isolation of disadvantaged groups; and (iv) combat social exclusion (INATEL, 2018). This programme is based on a rationale of positive discrimination, favouring participants with low income. Therefore, the amount payable by each participant varies according to his/her income. Thus, four groups were created based on the monthly income of participants. In the first group (monthly income up to €438.81 inclusive) the tourism trips are free.

Table 4.1 presents a description of the social tourism programmes offered to seniors by the INATEL Foundation.

4.3.3 Demand and benefits of the social tourism programmes for seniors offered by the INATEL Foundation

The social tourism programmes offered by INATEL have had a very important role in stimulating Portuguese access to tourism activities for seniors. Without these programmes, a large number of Portuguese low-income seniors would not have had the opportunity to experience tourism activities. First, these programmes overcome seniors' financial constraints to participating in tourism activities, because the trips organized in the scope of these programmes are subsidized. Second, they facilitate the tourism planning process, since all arrangements related to the trips are undertaken by the INATEL Foundation. Moreover, the packages purchased by the seniors include, among other features, transport, accommodation, meals, professional assistance of a guide, medical services, insurance and entertainment.

An analysis of the number of participants in the STP for seniors described in the previous section (Fig. 4.1) clearly reveals heterogeneity in the number of participants between 1995 (launch date of the first STP for seniors) and 2019. During this period, three stages may be identified in terms of the evolution of INATEL's social tourism programmes for seniors, considering the number of participants in the programmes: (i) expansion (1995–2005); (ii) contraction (2006–2012); and (iii) stagnation (2013–2019). In 2020 the number of participants in these social tourism programmes significantly decreased as a consequence of the Covid-19 pandemic.

The first stage (1995–2005) was marked by a considerable continuous growth, with the exception of 2003, in the number of beneficiaries of the two social tourism programmes offered by the INATEL Foundation (STEP and HTSTP). The number of participants in this period had a global growth of 317%. The second stage started in 2006 and was characterized by a decrease in the number of beneficiaries of the social tourism programmes. In 2006, 54,832 seniors participated in the STEP and HTSTP, while in 2012 the number of participants in four STP (STEP, HTSTP, 60+A and PAH) was less than half of the number in 2006 (22,841 participants), with a global decrease of 58%. Finally, in the third stage identified (2013–2019), although there was an offer of various social tourism programmes (STEP, HTSTP, 60+A, QDP, 55+.pt), no considerable changes in the number of participants can be observed. The number of participants in 2013 in four STP was 6793 and in 2019 the number of participants in the Programme INATEL 55+.pt was 5445. Therefore, to reverse this negative trend it is of utmost relevance to intensify the offer of social tourism initiatives to the senior market. The relevance of these types of programmes in the future will be even greater due to the active ageing of the Portuguese population. According to EUROSTAT (2020), in 2019 the old-age dependency ratio in Portugal (population 65 and over to population 15 to 64 years) was 33.9%, while for 2050, projections suggest 62.8%.

Another important issue related to the relevance of the social tourism programmes offered by INATEL to the senior market is the identification of the benefits obtained by the seniors who participated in these programmes. Therefore, two empirical studies were undertaken, in the scope of two research projects carried out by a group of researchers from the University of Aveiro for the INATEL Foundation (Eusébio *et al.*, 2009a, b). In order to examine the benefits perceived by seniors, two questionnaires

Table 4.1. Description of the social tourism programmes offered by the INATEL Foundation, since 1995, to seniors.

Name of the social programme	Date of the first edition	Date of the last edition	Destinations	Duration of tourism trips	Components of tourism trips	Target group
Senior Tourism Programme (SETP) [*Programa Turismo Sénior*]	1995	2017	Portugal and Spain	8 days (7 nights)	• Transportation and accommodation with full board • Constant monitoring by a recreational assistant • Participation in cultural and entertainment activities • Insurance against personal accidents • Baggage handling service at check-in and check-out	Portuguese citizens aged 60 and over, accompanied by spouse, a son/daughter with a degree of disability not superior to 60% or by a citizen aged over 18 years
Health and Thermal Treatments Senior Tourism Programme (HTSTP) [*Programa Saúde e Termalismo Sénior*]	1997	2016	Portugal	15 days (14 nights)	• Transportation and accommodation with full board • Constant monitoring by a recreational assistant • Participation in cultural and entertainment activities • Insurance against personal accidents • Baggage handling service at check-in and check-out • Thermal treatment (not included in the price of the trip)	Portuguese citizens aged 60 years or older, who may be accompanied by their spouses, independently of their age and requirements of thermal treatments, or by another Portuguese citizen, aged over 18, if assistance is needed

Continued

Table 4.1. Continued

Name of the social programme	Date of the first edition	Date of the last edition	Destinations	Duration of tourism trips	Components of tourism trips	Target group
Programme 60+ Azores (60+A) [*Programa 60+ Açores*]	2008	2013	Nine islands of the Azores Archipelago	8 days (7 nights)	• Flight and accommodation with full board • Transfers • Constant monitoring by a recreational assistant • Cultural activities and visits to other tourism attractions • Insurance against personal accidents	Residents in the Azores Archipelago aged 60 or over, with low incomes and consequently more exposed to the phenomenon of isolation, loneliness and social exclusion, to permit their participation in tourism activities through the payment of a symbolic participation fee
Programme Always on Holiday (PAH) [*Programa Sempre Em Férias*]	2011	2012	Portugal	1 month minimum with no maximum	• Transport and accommodation with full board in the hotel units of the INATEL Foundation • Constant monitoring by a recreational assistant • Two cultural tourism tours • Participation in recreational activities • Insurance against personal accidents • Health care • Baggage handling service at check-in and check-out • Laundry service	Portuguese citizens aged 60 or over, autonomous and in full enjoyment of their physical and mental abilities

Continued

Table 4.1. Continued

Name of the social programme	Date of the first edition	Date of the last edition	Destinations	Duration of tourism trips	Components of tourism trips	Target group
Quiet Days Programme (QDP) [*Programa Dias Tranquilos*]	2013	2017	Portugal	15 days	• Lodging with full board • Laundry service	Portuguese citizens aged 55 or over who wish to benefit from holidays and rest in Hotel Units of the INATEL Foundation for a minimum stay of 15 days
INATEL programme 55+.pt (P55+) [Progama INATEL 55+.pt]	2018	(Ongoing)	Portugal	6 days (5 nights)	• Transports and accommodation with full board • Constant monitoring by a recreational assistant • Cultural tours • Participation in recreational activities • Regional lunches • Tasting of regional products • Musical entertainment at the hotel • Educational activities, including training in areas such as citizenship, digital literacy, health and well-being • Insurance against personal accidents	Portuguese citizens aged 55 or over at risk of social exclusion or with low income

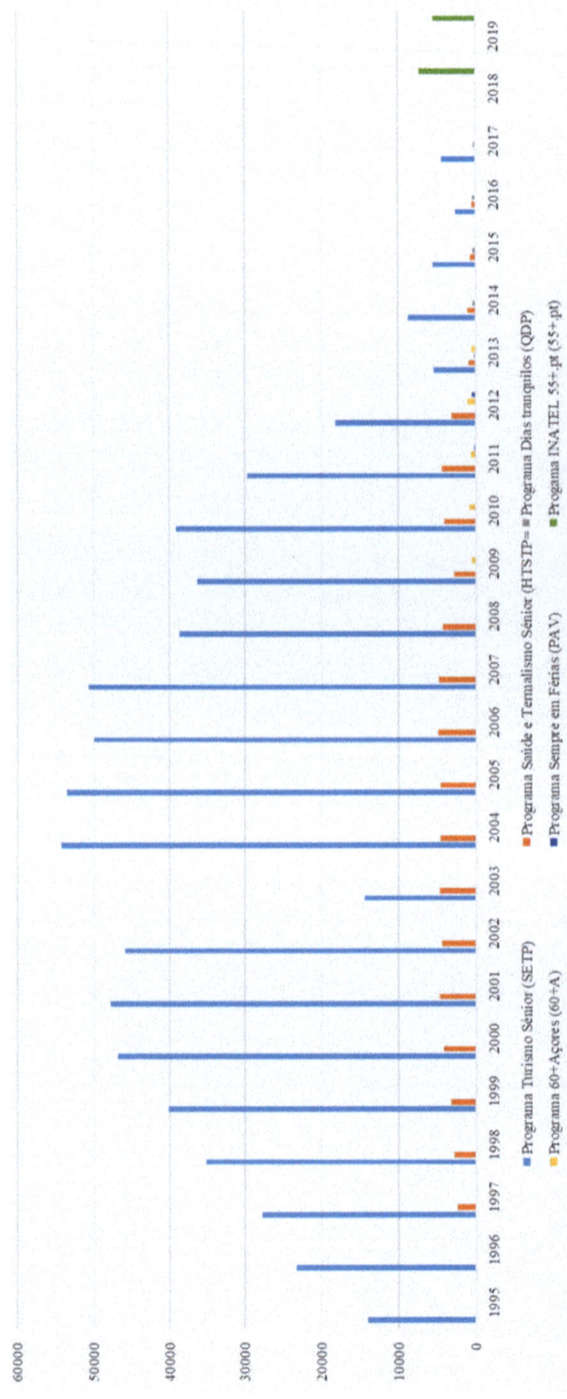

Social Tourism Programmes	1995	1996	1997	1998	1999	2000	2001	2002	2003	2004	2005	2006	2007	2008	2009	2010	2011	2012	2013	2014	2015	2016	2017	2018	2019
Programa Turismo Sénior	13948	23432	27860	35210	40159	46788	47785	45850	14427	54233	53480	49876	50569	38728	36329	39055	29710	18243	5414	8668	5580	2576	4381		
Programa Saúde e Termalismo Sénior			2408	2859	3187	4183	4685	4342	4659	4591	4613	4956	4809	4274	2807	4037	4386	3078	936	1027	675	462			
Programa 60+Açores															504	777	587	1025	444						
Programa Sempre em Férias																	184	495							
Programa Dias tranquilos																			219	354	345	330	285		
Progama INATEL 55+.pt																								7250	5445

Fig. 4.1. Number of participants in the social tourism programme offered by the INATEL Foundation for the senior market. (Based on Eusébio *et al.*, 2009a, b, Eusébio *et al.*, 2013, and INATEL, 2010, 2011, 2012, 2013, 2014, 2015, 2016, 2017, 2018, 2019a.)

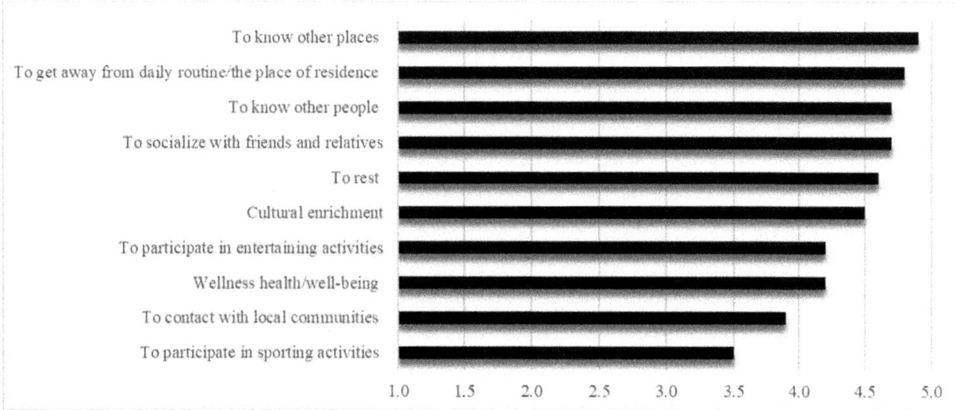

Fig. 4.2. Benefits obtained by participants in SETP. Scale from 1 'No benefits obtained' to 5 'Great benefits'.

(Adapted from Eusébio *et al*., 2009a, b, Eusébio *et al*., 2013.)

were administered to participants in the STEP (between 2001 and 2005) and in the HTSTP (between 2000 and 2007). A total of 1260 seniors participated in this survey (1000 regarding the SETP and 260 the HTSTP). To examine the benefits obtained, seniors were asked to indicate their opinion concerning 10 items by using a Likert-type scale from 1 (no benefit obtained) to 5 (great benefits obtained) (Figs 4.2 and 4.3).

The low frequency with which seniors report having taken holidays away from home before participating in the social tourism programmes analysed (SETP and HTSTP) highlights the importance of these programmes in making tourism accessible for all. Before participating in the two social tourism programmes, about 37% of SETP respondents had never or seldom taken holidays and about 62% of HTSTP participants had never or seldom received thermal treatments (Eusébio *et al*., 2009a, b). The main reasons mentioned by the respondents for never or seldom taking holidays are financial constraints (reported by 54% of SETP participants and 42% of the HTSTP participants) (Eusébio *et al*., 2009a, b, Eusébio *et al*., 2013, 2017).

Concerning the benefits obtained by seniors from their participation in SETP and HTSTP, presented in Figs. 4.2 and 4.3, seniors reported

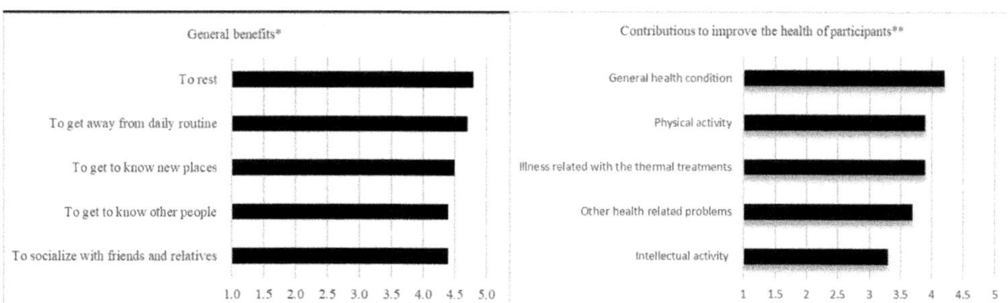

Fig. 4.3. Benefits obtained by participants in HTSTP.

* scale from 1 'No benefits obtained' to 5 'Great benefits'

** scale from 1 'Did not contribute anything' to 5 'Contributed a lot'.

(Adapted from Eusébio *et al*., 2009a, b, Eusébio *et al*., 2013.)

the main benefits of the STEP as being: the opportunity to visit other places, getting away from daily routine, getting to know other people, socializing with friends and relatives, and resting, as well as cultural enrichment (Fig. 4.2). Despite the relevance of relaxation, senior participants of SETP also recognize the value of participating in entertainment activities and opportunities to be in contact with the population of the local destination and to improve health and well-being.

The HTSTP also contributes greatly to improving seniors' health (Fig. 4.3). The respondents mentioned that their general health condition, the particular health condition associated with the thermal treatment and physical activity improved considerably with their participation in this programme. Improvements regarding seniors' intellectual activity were less perceived than health benefits (Fig. 4.3). Additionally, some seniors interviewed also mentioned that their health-related costs decreased after participation in the HTSTP. About 18% of respondents mentioned a decrease in medical consultation costs, 25% in physiotherapy costs and 35% in medication expenses (Eusébio *et al.*, 2009b). These findings concerning the benefits perceived by participants in the two social tourism programmes offered by INATEL to seniors (SETP and HTSTP) clearly show that these kinds of social initiatives greatly contribute to increasing seniors' participation in tourism and that the seniors obtained substantial benefits from their participation, corroborating the findings of previous research (e.g. Smith and Hughes, 1999; Minnaert *et al.*, 2009; McCabe and Johnson, 2013). Moreover, these results also highlight the importance of developing and financing this kind of social tourism programme.

4.4 Conclusions and Implications

Social tourism initiatives have proven to be relevant in stimulating less advantaged citizens' participation in tourism (Minnaert *et al.*, 2009), with diverse social tourism models observable in Europe (EESC, 2006). The present study presents the Portuguese model of social tourism supply, referring to the main public institution providing corresponding initiatives, the INATEL Foundation, its evolution and programmes targeting the country's senior population, particularly those with low-income conditions. Also data showing the participation of seniors in these programmes over the past 25 years are analysed, suggesting a relative decrease and the need to stimulate participation among seniors, given the many benefits related to these initiatives, as identified in several studies (Moal-Ulvoas and Taylor, 2014; Morgan *et al.*, 2015; Sedgley *et al.*, 2018; Hwang and Lee, 2019).

The results of an empirical study addressing participants in two INATEL social tourism programmes targeted at senior citizens, presented in this chapter, highlight the multiple and significant benefits that participants derived from their experience. These are mainly associated with features that may be linked to the notion of (physical, cognitive and social) 'active ageing' – namely, getting to know new places and people, socializing, cultural enrichment and physical activities – and to corresponding benefits regarding both health and social inclusion of these sometimes-isolated people, thereby enhancing their quality of life. General benefits associated with relaxation, escaping from one's routine, socialization and cultural enrichment may be considered important psychological benefits that further improve the participants' health. The specific health treatments in the thermal programme enhance the well-being of the elderly and reduce the health-related costs supported by the participants after the programme ends.

A major contribution of INATEL programmes is that they permit the participation of those frequently excluded from access to leisure and tourism owing to financial constraints. The large set of benefits that seniors attributed to their participation in INATEL programmes show their success and justify the corresponding investments, while participation seems to need further stimulation among the elderly to reverse the apparent negative trend in participation.

Tourism is, in fact, frequently recognized as a source of well-being and quality of life (Kim *et al.*, 2015). It is, according to Dann (2002), especially relevant for the elderly, who often have more time at their disposal to engage in tourism and reveal particular needs due to their frequently more fragile health and social isolation, which may be addressed through participation in tourism, but who unfortunately suffer from financial constraints (Fleischer and Pizam, 2002; Alén *et al.*, 2017; EUROSTAT, 2019b). Social tourism initiatives may help overcome these barriers (Medarić *et al.*, 2016), as also observable in the survey results presented here and may thereby permit an improvement to be achieved in the overall living conditions of senior citizens, producing significant health, social and economic benefits (Eusébio *et al.*, 2013).

References

Alén, E., Losada, N. and de Carlos, P. (2017) Profiling the segments of senior tourists throughout motivation and travel characteristics. *Current Issues in Tourism* 20(14), 1454–1469. DOI: 10.1080/13683500.2015.1007927.

Carneiro, M.J., Eusébio, C., Kastenholz, E. and Alvelos, H. (2013) Motivations to participate in social tourism programmes: a segmentation analysis of the senior market. *Anatolia* 24(3), 352–366. DOI: 10.1080/13032917.2013.767212.

Cisneros-Martínez, J.D., McCabe, S. and Fernández-Morales, A. (2018) The contribution of social tourism to sustainable tourism: a case study of seasonally adjusted programmes in Spain. *Journal of Sustainable Tourism* 26(1), 85–107. DOI: 10.1080/09669582.2017.1319844.

Daniels, M.J., Harmon, L.K., Brayley, R.E., Vese, R. and Park, M. (2019) Honor flight: recognition of preeminent-mature tourists. *Annals of Tourism Research* 74, 95–107. DOI: 10.1016/j.annals.2018.10.006.

Dann, G.M.S. (2002) Senior tourism and quality of life. *Journal of Hospitality & Leisure Marketing* 9(1–2), 5–19. DOI: 10.1300/J150v09n01_02.

EESC (2006) Opinion of the European economic and social Committee on social tourism in Europe. *Official Journal of the European Union* 23 December 2006, 67–77.

European Commission (EC) (2010) *Calypso Study on Social Tourism Portugal*. Ref. Ares(2014)3236031 - 01/10/2014.

EUROSTAT (2019a) EUROSTAT data base. Available at: https://ec.europa.eu/eurostat/data/database (accessed 25 February 2019).

EUROSTAT (2019b) EUROSTAT statistics explained – tourism trends and ageing. Available at: https://ec.europa.eu/eurostat/statistics-explained/index.php/Tourism_trends_and_ageing (accessed 26 February 2019).

EUROSTAT (2020) Demographic balances and indicators by type of projection. Available at: https://appsso.eurostat.ec.europa.eu/nui/show.do?dataset=proj_19ndbi&lang=en (accessed 20 July 2020).

Eusébio, C., Costa, C., Carneiro, M.J., Kastenholz, E. and Alvelos, H. (2009a) *Estudo de avaliação dos impactos sócio-económicos dos Programas de Turismo Sénior de 2001 a 2005 [Study to assess the socioeconomic impacts of the Senior Tourism Programs from 2001 to 2005]*. INATEL Foundation.

Eusébio, C., Carneiro, M.J., Kastenholz, E. and Alvelos, H. (2009b) *Estudo de avaliação dos impactos sócio-económicos dos Programas Saúde e Termalismo Sénior de 1997 a 2007 [Study to assess the socioeconomic impacts of the Health and Thermal Senior Tourism Programs from 1997 to 2007]*. INATEL Foundation.

Eusébio, C., Carneiro, M.J., Kastenholz, E. and Alvelos, H. (2013) The socioeconomic impacts of INATEL social tourism programs for the senior market. In: Diekmann, A. and Jolin, L. (eds) *Social Tourism: International Perspectives the Contribution of Research*. Presses de l' Université du Québec, Québec, Canada, pp. 197–213.

Eusébio, C., Carneiro, M.J., Kastenholz, E. and Alvelos, H. (2016) The impact of social tourism for seniors on the economic development of tourism destinations. *European Journal of Tourism Research* 12, 5–24.

Eusébio, C., Carneiro, M.J., Kastenholz, E. and Alvelos, H. (2017) Social tourism programmes for the senior market: a benefit segmentation analysis. *Journal of Tourism and Cultural Change* 15(1), 59–79. DOI: 10.1080/14766825.2015.1117093.

Fleischer, A. and Pizam, A. (2002) Tourism constraints among Israeli seniors. *Annals of Tourism Research* 29(1), 106–123. DOI: 10.1016/S0160-7383(01)00026-3.

Huang, L. and Tsai, H.-T. (2003) The study of senior traveler behavior in Taiwan. *Tourism Management* 24(5), 561–574. DOI: 10.1016/S0261-5177(03)00008-6.

Huber, D., Milne, S. and Hyde, K.F. (2018) Constraints and facilitators for senior tourism. *Tourism Management Perspectives* 27, 55–67. DOI: 10.1016/j.tmp.2018.04.003.
Hwang, J. and Lee, J.J. (2019) A strategy for enhancing senior tourists' well-being perception: focusing on the experience economy. *Journal of Travel & Tourism Marketing* 36(3), 314–329. DOI: 10.1080/10548408.2018.1541776.
IMSERSO (2019a) Programa de Turismo del Imserso. Available at: www.imserso.es/imserso_01/envejecimiento_activo/vacaciones/servicios/index.htm (accessed 25 February 2019).
IMSERSO (2019b) Programa de Termalismo del Imserso. Available at: www.imserso.es/imserso_01/envejecimiento_activo/termalismo/index.htm (accessed 25 February 2019).
INATEL (2010) *Relatórios e Contas 2010*. INATEL, Lisbon.
INATEL (2011) *Relatório de Contas 2011*. INATEL, Lisbon.
INATEL (2012) *Relatório de Contas 2012*. INATEL, Lisbon.
INATEL (2013) *Relatório de Contas 2013*. INATEL, Lisbon.
INATEL (2014) *Relatório de Contas 2014*. INATEL, Lisbon.
INATEL (2015) *Relatório de Contas 2015*. INATEL, Lisbon.
INATEL (2016) *Relatório de Contas 2016*. INATEL, Lisbon.
INATEL (2017) *Relatórios e Contas 2017*. INATEL, Lisbon.
INATEL (2018) *Relatórios e Contas 2018*. INATEL, Lisbon.
INATEL (2019a) *Relatórios e Contas 2019*. INATEL, Lisbon.
INATEL (2019b) Programa INATEL 55+.pt: Regulamento específico [Programme INATEL 55+. pt: Specific regulation]. Available at: www.inatel.pt/getmedia/f42506c6-a599-4dbf-9a2d-2cf06b569c72/Regulamento_Especifico.aspx (accessed 25 February 2019).
INATEL (2019c) INATEL 55+. Available at: www.inatel.pt/Fundacao/Para-onde-ir/INATEL-55-pt.aspx (accessed 25 February 2019).
Kazeminia, A., Del Chiappa, G. and Jafari, J. (2015) Seniors' travel constraints and their coping strategies. *Journal of Travel Research* 54(1), 80–93. DOI: 10.1177/0047287513506290.
Kim, H., Woo, E. and Uysal, M. (2015) Tourism experience and quality of life among elderly tourists. *Tourism Management* 46, 465–476. DOI: 10.1016/j.tourman.2014.08.002.
Lee, S.H. and Tideswell, C. (2005) Understanding attitudes towards leisure travel and the constraints faced by senior Koreans. *Journal of Vacation Marketing* 11(3), 249–263. DOI: 10.1177/1356766705055716.
McCabe, S. and Johnson, S. (2013) The Happiness factor in tourism: subjective well-being and social tourism. *Annals of Tourism Research* 41, 42–65. DOI: 10.1016/j.annals.2012.12.001.
Medarić, Z., Gabruč, J. and Sedmak, M. (2016) Social tourism benefits for seniors. *Academica Turistica* 9(2), 113–116.
Minnaert, L., Maitland, R. and Millher, G. (2009) Tourism and social policy: the value of social tourism. *Annals of Tourism Research* 36(2), 316–334.
Minnaert, L., Diekmann, A. and McCabe, S. (2012) Defining social tourism and its historical context. In: MacCabe, S., Minnaert, L. and Diekmann, A. (eds) *Social Tourism in Europe: Theory and Practice*. Channel View Publications, Bristol, UK, pp. 18–.30.
Moal-Ulvoas, G. and Taylor, V.A. (2014) The spiritual benefits of travel for senior tourists. *Journal of Consumer Behaviour* 13(6), 453–462. DOI: 10.1002/cb.1495.
Morgan, N., Pritchard, A. and Sedgley, D. (2015) Social tourism and well-being in later life. *Annals of Tourism Research* 52, 1–15. DOI: 10.1016/j.annals.2015.02.015.
Otoo, F.E. and Kim, S.S. (2020) Analysis of studies on the travel motivations of senior tourists from 1980 to 2017: progress and future directions. *Current Issues in Tourism* 23(4), 393–417. DOI: 10.1080/13683500.2018.1540560.
Pestana, M.H., Parreira, A. and Moutinho, L. (2020) Motivations, emotions and satisfaction: the keys to a tourism destination choice. *Journal of Destination Marketing & Management* 16, 100332. DOI: 10.1016/j.jdmm.2018.12.006.
Pezeshki, F., Ardekani, S.S., Khodadadi, M., Alhosseini Almodarresi, S.M. and Hosseini, F.S. (2019) Cognitive structures of Iranian senior tourists towards domestic tourism destinations: a means-end chain approach. *Journal of Hospitality and Tourism Management* 39, 9–19. DOI: 10.1016/j.jhtm.2019.01.008.
Pine, B.J. and Gilmore, J.H. (1998) Welcome to the experience economy. *Harvard Business Review* 76(4), 97–105.

Sedgley, D., Haven-Tang, C. and Espeso-Molinero, P. (2018) Social tourism & older people: the IMSERSO initiative. *Journal of Policy Research in Tourism, Leisure and Events* 10(3), 286–304. DOI: 10.1080/19407963.2018.1465064.

Shavanddasht, M. (2018) Grandparent's segmentation by the tourism motivation: travelling with or without grandchildren. *Young Consumers* 19(2), 141–158. DOI: 10.1108/YC-07-2017-00719.

Small, J. (2003) The voices of older women tourists. *Tourism Recreation Research* 28(2), 31–39. DOI: 10.1080/02508281.2003.11081402.

Smith, V. and Hughes, H. (1999) Disadvantaged families and the meaning of the holiday. *International Journal of Tourism Research* 1(2), 123–133. DOI: 10.1002/(SICI)1522-1970(199903/04)1:2<123::AID-JTR146>3.0.CO;2-R.

United Nations (1948) *General Assembly Resolution 217 A*. UN, Paris.

UNWTO (2019) Global code of ethics for tourism. Available at: http://ethics.unwto.org/en/content/global-code-ethics-tourism (accessed 1 March 2019).

UNWTO (2020) INATEL Foundation. Available at: https://www.unwto.org/affiliate-member-organization/18943 (accessed 19 July 2020).

5 Social Tourism Programmes for Seniors: The Case Study of Slovenia

Janja Gabruč[1]* and Zorana Medarić[2]
[1]Faculty of Tourism Studies, University of Primorska, Koper, Slovenia; [2]Faculty of Tourism Studies, University of Primorska and Science and Research Centre, Koper, Slovenia

5.1 Introduction

Despite a long tradition of different tourism programmes in Slovenia recognized as social tourism (e.g. trade union tourism programmes), this form of tourism has yet to emerge as a topic in public discussion or in research. In recent years, Slovenia has increasingly begun to identify the problems of social tourism and the opportunities for its development, not least by its funding of the project, 'Identification, classification and potential for the development of social tourism products in Slovenia, with an emphasis on programmes for senior citizens' (MGRT, 2017), which was supported by the Slovenian Research Agency and the Ministry of Economic Development and Technology, and is the basis for the present chapter. The project was run between 2015 and 2017 by the Faculty of Tourism Studies of the University of Primorska, in cooperation with the Science and Research Centre Koper and the Faculty of Design (associate member of the University of Primorska).

It has also increasingly been recognized that seniors (over the age of 65) represent an important potential tourism market. Seniors currently constitute one fifth of the population of the EU (EUROSTAT, 2018). In Slovenia, the share of the senior population is also increasing, representing 16.5% of the total population in 2010, and increasing to 19.1% at the beginning of 2017 (SURS, 2018). Seniors therefore represent an important part of society facing issues related to quality of life and their right to quality leisure time, including tourism activities. Many seniors are unable to take part in tourism activities owing to their low socioeconomic status or other conditions such as poor health, mobility problems and various types of disability. According to the Statistical Office of the Republic of Slovenia, 64.34% of citizens older than 65 did not go on private trips during 2017, frequently citing reasons of ill-health or financial circumstances (SURS, 2018).

In order to understand the issue of social tourism in Slovenia, it is necessary to present the wider context and look at the past. Diverse programmes for social tourism have a long tradition in Slovenia, particularly youth holidays and holidays for employees and their family members within the framework of trade union tourism.

Youth holidays in Slovenia have been organized since the 1950s: organizations such as the Association of the Friends of Youth, Scouts, Red Cross, Fire Fighting Associations, trade unions and others started by organizing holidays

*Corresponding author: Janja.Gabruc@fts.upr.si

 Social Tourism: Global Challenges and Approaches (eds J. Lima and C. Eusébio)
DOI: 10.1079/9781789241211.0006

in the form of camps and health colonies on the coast or in the hills, as well as abroad (Austria), in tents, school buildings, gymnasiums, fire stations, mountain huts and trade union resorts (Struc *et al.*, 2013). The Association of Friends of Youth Slovenia (ZPMS) remains the largest organizer of youth holidays in Slovenia, having continuously provided holidays as one of their core activities since their very beginning, with more than 60 years of tradition. From the outset, the purpose of these holidays has been to focus on providing benefits for children, such as health, education and recreation. Nowadays the organization is committed to providing children with a high-quality leisure-time environment and to encourage creativity, activity, and the discovery and expression of their talents. From the start, these programmes were known for their social focus, ensuring that all children, regardless of their social status, could take part in such holidays; children from socially disadvantaged families received priority referrals, and the organization's policy was that contributions to socially vulnerable parents were appropriate to their economic circumstances. Compared with the period prior to Slovenian independence, the share of children taking part in holidays is now lower; according to the ZPMS, approximately 20,000 children per year currently take part, while at its peak in 1959 there were around 35,000 Slovenian children (Gabruč, 2014a).

The same applies to another once-widespread example of social tourism – trade union tourism in Slovenia – which today, however, is in decline. Social tourism in the post-war period defined two main aspects: (i) cash benefits for employees and their families, lower travel and accommodation charges for retired workers, and holiday allowances; and (ii) accommodation facilities in holiday resorts and company-owned accommodation (Župančić and Puljić, 2017). Until 1990, the main players and initiators of social tourism in Slovenia were trade unions, but today it is mainly companies that (still) enable employees and their family members to take more affordable holidays. According to union sources (Leban, 2009), this form of tourism was highly developed in the former Yugoslavia prior to the 1990s, with over 153,000 beds in homes, campsites, caravans, apartments, bungalows and small company-owned hotels. The capacities were located primarily on the Croatian coast in the former Yugoslavia, and now fall inside Croatia itself. Some of this capacity was in the mountains and by lakes, as this holiday segment was very popular. The vast majority of companies had their own facilities where employees could spend their holidays with families at a lower cost. From 1990, this segment of tourism changed significantly and decreased. It is estimated that there are approximately 15,000 beds still in use by trade union tourism in Slovenia (Leban, 2009).

In the past, seniors mostly spent their holidays in employers' accommodation, as part of their union's holiday capacity or their entitlement after taking retirement. The right to use such tourist capacity and the entitlement to receive a holiday allowance did not end with retirement. On the contrary, retirees were often involved in a variety of jobs at these holiday resorts during the high season, working on infrastructure maintenance, as waiters or cooks, supporting longer and 'cost-free' holidays. As previously stated, only a small part of trade union tourism has been preserved to date, although, curiously, the holiday allowance has remained, with all workers and pensioners receiving it – albeit significantly lower in value nowadays. It is not often used for that purpose, but is instead used to improve their poor socioeconomic situation (P.R. Trade Union of Catering and Tourism representative, Ljubljana, 2016). The holiday allowance is also one of the rare mechanisms of (social) tourism that can be understood as systemic and, with dedicated use, could be used to achieve significant progress in the development of social tourism. Some companies have held on to, or have started to reinvest in their holiday facilities, primarily for their employees and retirees to use. At the same time, these social infrastructure providers are increasingly commercializing and freeing up capacity to outside guests, mainly for economic reasons (Sindikalni počitniški domovi so spet zanimivi, 2017). These days, the primary consideration for seniors' quality of life and their participation in tourism is maintained by local pensioner associations that are affiliated to the Association of Pensioners' Associations of Slovenia (ZDUS). Tourism represents an important part of such associations' activities, in which, in cooperation with selected tourism companies, short, mostly 1-day excursions, short seaside trips and holidays are prepared for their members. Senior citizens enjoy taking part in these programmes and are satisfied with them, but owing to poor socioeconomic circumstances, many Slovenian pensioners cannot afford such trips and holidays (MGRT, 2017; SURS, 2018). In such situations, a number of humanitarian organizations assist socially disadvantaged seniors by obtaining the necessary funding for holidays through a variety of national campaigns.

The aim of this chapter is to present social tourism programmes for seniors in Slovenia by focusing on three aspects: (i) What programmes, identifiable within the framework of social tourism, exist in Slovenia? (ii) What adjustments do they include to enable seniors to take part? And (iii) what are the benefits of such programmes?

The programmes were analysed in two ways: (i) through the analysis of secondary sources, i.e. review of programme webpages and various data provided by programme organizers in the form of reports; and (ii) through a qualitative survey of key stakeholders in the field of social tourism for seniors. Qualitative research was conducted (semi-structured interviews took place between January and March 2016) with different players engaging with social tourism for seniors in Slovenia, including the Red Cross, the Premiki Institute and Agency for Accessible Tourism, the Tourism and Hospitality Trade Union of Slovenia, ZDUS, several (local) pensioners' associations, Relax travel agency, Hotel Delfin and the Paraplegics Association of Slovenia. Interview questions focused on various topics, such as their views on the development of social tourism programmes and research into existing social tourism programmes in Slovenia, the beneficiaries, extent and duration of programmes, organization and financing of programmes, as well as benefits for seniors.

The following sections of the chapter reflect on each one of these questions.

5.2 Characteristics of Social Tourism Programmes for Seniors in Slovenia

In this section, some features of the initiatives and programmes identified will be presented and placed within existing theoretical models. As a starting point for these considerations of social tourism, we have used the models of social tourism by Minnaert *et al.* (2013), which include four models, or interpretations, of social tourism programmes: participation, inclusion, adaptation and stimulation models, which will be presented later.

Social tourism systems and programmes vary from country to country and can be interpreted in various ways. According to McCabe *et al.* (2012), social tourism can include commercial and/or non-commercial, governmental and/or private organizations, each having a distinct understanding of what social tourism means. Within the field of social tourism in Slovenia, there are various initiatives that can be divided into the following four groups:

1. Trade union initiatives.
2. Initiatives of charitable, non-governmental, non-profit organizations.
3. Social entrepreneurship initiatives.
4. Private initiatives (market-oriented initiatives).

In the first category, for example, can be found the initiative of the Trade Union for Catering and Tourism, which, as a one-stop shop, collects offers of trade union accommodation from companies in different sectors in Slovenia and Croatia, and thus addresses a very wide circle of people, not just members of the respective trade union. Among the initiatives from non-governmental or non-profit organizations, various humanitarian organizations such as the Red Cross and Caritas, which organize holidays for socially vulnerable seniors, were considered. In the third group were placed the initiatives of the Premiki Institute and Agency for Accessible Tourism, which specializes in accessible tourism with services that are adjusted to the needs of seniors. As private initiatives, services from market-oriented travel agencies were considered, in which seniors are recognized as an important market segment and the offers are adjusted to the needs of seniors, often connecting with pensioners' associations (Table 5.1).

From a *content perspective*, tourist programmes intended for seniors are very diverse. Tourist agency programmes aim to offer week-long holidays, most often to the Croatian coast, as well as to spas in Slovenia and increasingly

Table 5.1. Overview of initiatives and social tourism programmes in Slovenia. (Based on MGRT, 2017.)

Provider	Programme	User	Source of funding
Slovenian Red Cross	'Let's take them to the sea' National humanitarian campaign.	Seniors (and children)	Humanitarian funds
Trade Union of Catering and Tourism of Slovenia	Portal: sindikalne-pocitnice.si	Seniors (and others)	Participants' contributions
Slovenian Federation of Pensioners' Associations: Hotel Delfin	Holidays for seniors	Seniors	Participants' contributions
Slovenian Paraplegic Association; The Pacug Health Resort Centre	Social and follow-up rehabilitation Accessible accommodation/services	Paraplegic seniors and other paraplegics Seniors and others	Public funds Participants' contributions
Premiki – Institute and Agency for Accessible Tourism	Customized/adjusted tourism, offers for people with special needs and their relatives/travel companions	Disabled seniors and other disabled people	Participants' contributions
Local pensioner organizations and clubs in cooperation with travel agencies (e.g. Relax travel agency)	Customized tourist offers for seniors, e.g. 'Senior Club'	Seniors	Participants' contributions

to Bosnia; day trips, such as seaside excursions; and package holidays – 2 or 3 days – by coach in their locality. Agencies adjust to the wishes and needs of pensioner associations in price and organization and often prepare joint offers. Humanitarian holidays, such as the Red Cross programme at Debeli Rtič, are usually weekly and provide comprehensive assurance of accommodation, food, entertainment and health activities (e.g. exercise, therapy, check-ups). Most of the interviewees stressed the importance of comprehensive offers, including transport, food and attendants: all-inclusive packages, where everything is organized, are highly popular.

Seniors are one of the four main social tourism target groups in the European context (Diekmann *et al.*, 2012). However, the beneficiaries of social tourism programmes for seniors differ from one another according to their socioeconomic circumstances and medical condition, but can roughly be divided into the following groups:

1. Seniors with a low socioeconomic status.
2. Seniors who require special infrastructure adjustments.
3. Seniors, as one of the target groups of tourists for whom special offers have been created.
4. Seniors, as only one of several target groups of tourists.

The first group, comprising economically deprived seniors, can be placed among the beneficiaries of the Red Cross humanitarian programme 'Let's take them to the sea', in which holidays are promoted by local Centres for Social Work as a means of preventative care and promoting general health. The second group of beneficiaries represents those seniors who require special infrastructure adjustments, such as disabled seniors using adjusted/accessible holiday facilities, and adjusted transport – for example, holiday facilities of the Slovenian Paraplegic Association. The third group of beneficiaries comprises those senior travellers for whom special offers have been created and tailored in terms of price, payment terms, content and organized activities (e.g. Relax travel agency in association with the (local) pensioners' associations). The fourth group of beneficiaries represents seniors as only one of several target groups of tourists who are consumers of standard tourism products – without special adjustments, for example, using only union-maintained – that is, affordable – holiday facilities (e.g. GIT Trade Union).

According to Diekmann and McCabe (2013), for social tourism to work there needs to be a system to deliver funds, and in this regard, two major types of funding can be observed: infrastructure-oriented funding or beneficiary-oriented funding. There is no systematic financing in Slovenia, with programme funding being obtained through a variety of channels; it should be emphasized that most programmes are financed from the beneficiaries' own resources, and that funding is generally beneficiary-oriented and to a lesser extent infrastructure-oriented. The sources of financing for social programmes are very diverse, with four financing channels identified:

- public sources of financing (Foundation for Financing Disabled and Humanitarian Organisations (FIHO), Health Insurance Institute of Slovenia (ZZZS), Ministry of Labour, Family, Social Affairs and Equal Opportunities (MDDSD), municipal contributions);
- humanitarian funding (from humanitarian campaigns and individual donations and sponsorship);
- organizations'/providers' own sources (e.g. membership fees); and
- beneficiaries' own sources (e.g. [state] holiday allowance).

Humanitarian programmes obtain funding from a variety of sources, such as FIHO, voluntary contributions and humanitarian actions. Similarly, health care programmes and programmes for rehabilitation that are intended for seniors with disabilities or specific medical conditions are also funded by FIHO, the ZZZS and municipal funds. As a rule, seniors pay for commercial programmes from their own funds, but pensioners' associations acting as mediators often negotiate low-cost programmes. The commercial programmes of tourist agencies and trade union accommodation are usually self-financing.

As previously stated, we have taken theoretical models of social tourism as the starting point for our consideration of social tourism, including four models, or interpretative aspects of social and tourism programmes: participation, inclusion, adaptation and stimulation models (Minnaert *et al.*, 2013).

Participation and adaptation models encourage participation in tourism, especially in groups that have greater difficulty accessing tourism as a result of health or financial circumstances, comprising young people, senior citizens, families and the disabled (European Commission, 2010a), while inclusion and stimulation models encourage broader society according to a 'tourism for all' principle. Inclusion models promote visitor-related initiatives and stimulation models promote host-related initiatives, where (social) tourism providers are willing to adjust their offers during the low season and alter them to the different groups of beneficiaries. The products offered are standard in respect of the participation and inclusion initiatives of social tourism, while the adaptation and stimulation models seek to bring the product closer to the specific/special needs of the beneficiaries involved, through infrastructure and other adjustments.

Models vary depending on who the beneficiaries of such initiatives are and on the characteristics of tourism products, which can be either completely standard or adjusted to the specific needs of beneficiaries and include certain 'specific provisions'. As one of the key elements that, in addition to the beneficiaries and the included infrastructure, should be considered in the analysis of social tourism systems, Diekmann and McCabe (2013) highlight that funding must be given special attention. In respect of the four previously mentioned social tourism models (relating to products and consumers), we consider the initiatives that can be placed within each model (Table 5.2).

Among the initiatives analysed, the Red Cross's 'Let's take them to the sea' can be placed within the framework of the participation model of social tourism, where standard tourism products are intended for social tourism users. This is a humanitarian campaign to raise funds for holidaying seniors who are selected on the recommendation of local Centres for Social Work. In the context of this campaign they are taken to a health resort at Debeli rtič, where they participate in a high-quality, well-designed health and social programme. The programmes and

Table 5.2. Social tourism models and related examples of Slovenian (social) tourism programmes for seniors.(Based on MGRT, 2017.)

Social tourism models (Minnaert *et al*., 2013)	(Social) tourism programmes for seniors – examples in Slovenia
Participation model	National humanitarian tourism programme for seniors (and children) 'Let's take them to the sea' organized by the Red Cross Slovenia
Inclusion model	Trade union tourism programmes for seniors organized by the Trade Union of Catering and Tourism of Slovenia ('sindikalne-pocitnice.si' web portal)
Adaptation model	Tourism programmes for individuals with special needs and their relatives, elderly, the physically handicapped and others, organized by the Premiki Institute and Agency for Accessible Tourism Tourism programmes specially adjusted to seniors, organized by the Slovenian Federation of Pensioners' Associations (Hotel Delfin) Social and follow-up rehabilitation and tourism programmes for physically handicapped people, including seniors, organized by the Slovenian Paraplegic Association (The Pacug Health Resort Centre)
Stimulation model	Off-season tourism programmes for seniors organized by pensioners' organizations/clubs in cooperation with travel agencies (e.g. Relax tourist agency and its Senior Clubs)

infrastructure are fully tailored to seniors. The key objectives of such programmes are the engagement of seniors in tourism, their social inclusion and enhancing their quality of life, as they are often isolated and unable to participate actively in tourism due to poor socioeconomic circumstances or social life. The programme involves between 200 and 300 people each year who would not otherwise be able to afford a vacation/treatment programme. However, self-employed pensioners may also participate in holidays and treatments (P.R. Red Cross representative, Ankaran, 2016, personal communication).

According to Gabruč (2014b), although comprehensive and inclusive social tourism practices addressing the widest possible group of people in Slovenia are unfamiliar, the example of trade union tourism is mentioned as the most closely related practice, and we can therefore place it as an inclusive model. Trade union and other pensioners can use the 'sindikalne-pocitnice.si' web portal to access cheaper holidays. Where holidaying capacity is not taken up by members, it becomes available to the general market, and spas (e.g. Banovci, Ptuj, Čatež, Moravske toplice), in particular, are widely visited by pensioners. One of the tasks of the trade union that established the portal is to develop future social tourism, in part to provide senior care and, to a greater extent, with the intention of filling their vacation capacities with the categories of tourists that are unrelated to holidaying: the elderly, the disabled, the socially vulnerable and the young (P.R. Trade Union of Catering and Tourism representative, Ljubljana, 2016, personal communication).

Among the companies analysed, the practices of the Premiki Institute and Agency for Accessible Tourism can be placed in the adaptation model of social tourism. Although their services are directed primarily at individuals with special needs and their relatives, they also adjust tourist services for the elderly, physically handicapped and others. Their website indicates which programmes are adjusted to specific groups (e.g. those with physical disabilities, with hearing disabilities, sight disabilities, etc.). Programmes for the elderly are adjusted in terms of transport, accessibility of the accommodation infrastructure, tourist attractions, ample viewing times and enriched programme content (P.R. institute representative, Ljubljana, 2016, personal communication). As a social enterprise, they focus on employing individuals from vulnerable groups and their programmes are not co-financed. In part, we can also place the programmes of Hotel Delfin, which is owned by the Union of Pensioners' Associations of Slovenia, into this model as it is oriented primarily towards the target group of seniors, and thus adjusts its programmes in terms of content and organized activities for older tourists. Their entire infrastructure is also adjusted to the elderly. They also organize free transfers for pensioners from all over Slovenia (P.R. hotel representative, Izola, 2016, personal communication). As an example of the adaptation model of social tourism, we can mention the programmes of the Pacug Health Resort Centre, which provides accommodation and other health-related services and affordable vacation/tourism products tailored to physically handicapped people for their members, quadriplegic and paraplegic visitors who come to them for restorative rehabilitation, as well as for other physically impaired guests. They are also aimed at seniors and other guests with mobility problems who need such adjustments, which, in addition to infrastructure, include interaction adjustments (P.R. health resort representative, Portorož, 2018, personal communication).

Though systems of (co-)funding for social tourism are unfamiliar in Slovenian practice (Gabruč, 2014a), it is common practice for tourism companies to provide specially designed off-season offers for the target group of seniors in cooperation with the pensioners' associations. These programmes can be placed within the stimulation model. The key features of such tourism products are reduced prices or (co)-financing for part of the services; for example, transport from different parts of Slovenia or content-tailored tourism products. According to a spokesperson of the pensioners' association (P.R. pensioners' association representatives, Ljubljana, 2016, personal communication), individual travel agencies recognize seniors as an important target group in tourism and regularly cooperate with pensioners' associations, discussing special discounts, content on offer, tailor-made transport, payment options for meals, organizing special escorts and so on. We can also place the Relax travel agency into this form of tourism (model); it has been organizing its specific senior programme for more than 15 years, arranging holidays before and after the high season for more than 20,000 Slovenian guests. The seniors themselves (both Slovene and wider) have been recognized as important yet unexploited potential and opportunity. Relax has cooperated with pensioners' associations for many years, negotiating special rates on affordable all-inclusive offers. They also specialize in senior clubs, by negotiating specific leasing of hotel capacity for elderly guests, with representatives and activities that are mainly adjusted to their needs (P.R. agency representative, Ljubljana, 2016, personal communication). This practice is also found in the Hotel Delfin, which organizes free transportation for pensioners (P.R. hotel representative, Izola, 2016, personal communication).

5.3 Support Mechanisms and Adjustments Providing Greater Access for Seniors

Adjustments to tourist programmes are usually discussed within the context of accessible tourism (Darcy, 2010; Bizjak *et al.*, 2011; Pagán, 2012; Domínguez Vila *et al.*, 2015). One extremely important role that concerns individuals – with or without special needs – is about reducing uncertainty in tourism (Minnaert, 2014), something that is of particular concern for people with little travel experience. Undoubtedly, one of the most important adjustments is the affordability of tourism products and systemic (public) financing of social or accessible tourism products (Diekmann and McCabe, 2011), which is difficult to imagine without the support of various financial mechanisms.

Through interviews and secondary analysis of seniors' programmes in Slovenia, we have identified a variety of support mechanisms and adjustments to social tourism programmes for the elderly; these have proven to be very important for them, and should be taken into account in establishing systemic policies for social tourism for seniors. The research has highlighted the importance of specific adjustments to these programmes, which enable greater access for seniors, mainly in their consideration of (i) (low) socioeconomic status; (ii) mobility/accessibility issues; and (iii) travel uncertainty. Recent work (Lee *et al.*, 2012; Morgan *et al.*, 2015; Nikitina and Vorontsova, 2015) shows more attention being paid to accessible tourism, including adjustments for seniors.

5.3.1 Socioeconomic status support mechanisms

Price is an extremely important factor for travel programmes for seniors; many of the elderly face social hardship, which can make it difficult for them to take part in tourism. Their social circumstances deteriorate, which is why the frequency of trips and holidays among pensioners drops. What adjustments, relating to the socioeconomic status of seniors, were identified through the analysis of social tourism programmes in Slovenia?

The worker's annual holiday allowance introduced at the time of the Yugoslavian federation is a systemic financial mechanism for all pensioners. Pensioners receive a holiday allowance from the Pension and Disability Insurance Institute each year. This is an annual allowance from contributions to their pension for the month of July to be used for vacationing. The amount of the pensioner's annual allowance depends on the size of their pension in the month that the allowance is paid and is received by all pensioners. The allowance for pensioners in 2018 is at least €90, and those with the lowest pensions receive €400. Although these are dedicated funds intended for annual leave, the allowance is often used for other purposes to make up for low incomes throughout the year, as noted in a statement by the spokesperson of ZDUS:

> In Slovenia, or, as it was in Yugoslavia, it is a tradition to receive this supplement, which is called the recreation or pensioner's allowance, received in the form of financial funding, which is basically awarded without purpose – this is what is said – but nobody checks how this is then spent. This is essentially a social corrective for people. Once you receive it, it is spent, for example, on wood for the winter, but in brief, it is not for holiday nor for recreation. (ZDUS spokesperson).

Tourism programme providers – pensioners' associations in cooperation with travel agencies – try to adjust tourism programmes for seniors with discounted prices (up to 30–40% cheaper for groups of pensioners), payment in instalments or free transfers. Many commercial providers of tourism services or products offer reduced pricing or special discounts for seniors. In this framework, we can also include the favourable payment terms offered by certain commercial providers, such as payment in several instalments, distributed over the year, working closely with pensioners' associations and adapting to their needs. Pensioners' associations often act as the intermediaries through which the programmes are organized and, because of the number of participants, can negotiate more favourable rates. Programmes are mainly implemented out of the high season, in spring and autumn, as prices are much lower than otherwise. This allows for the creation of favourable packages, organizing overnight stays in good hotels at low prices. At the end of May prices are even lower, with one-week all-inclusive holidays on the Adriatic available for €250–300. Travel agencies that have recognized seniors as a relevant target group, such as Relax, prepare travel or vacation programmes that are affordable for a wide range of older people, including adjusting the hotel category and the occupation dates outside the high season.

> More recently, I believe our packages are more available to a much wider population, and with some sort of helpful payment arrangements, such packages are more acceptable. That's why the number of guests is also growing... We are trying to make them accessible to everyone – through different programmes. This includes the hotel category, and, depending on the accommodation date, both in the Mediterranean and the Adriatic. We think that they are accessible to practically everyone, but unfortunately, we know that they cannot normally afford these holidays. (spokesperson for the Relax travel agency)

The next aspect of adaptation that enables integration of the elderly in tourism is (financial) integration into tourism for those with lower socioeconomic status. To a lesser extent, humanitarian programmes, such as the Red Cross's 'Let's take them to the sea' allows elderly and socially deprived seniors to go on vacation (around 250 individuals annually depending on the funds raised). Regional associations of the Red Cross of Slovenia cooperate with state Centres for Social Work by organizing holidays and providing funding for their implementation and, on the recommendation or opinion of those centres, then shortlist the seniors for such holidays. Financial resourcing is the main obstacle to the development of social tourism programmes, for which there is an increasing demand. As a result, the Red Cross receives part of these humanitarian funds from FIHO, and partly from donations.

In accordance with its annual programme, the ZZZS provides programmes for individuals suffering from serious conditions (muscular and neuromuscular diseases, paraplegia, cerebral palsy, along with the most severe form of generalized psoriasis, multiple sclerosis or paralysis with specific functional impairment), with the possibility of participating in groups and having adapted treatments given by specialists at appropriate clinics, institutions or other health care establishments. Rehabilitation participants have guaranteed co-financing for physiotherapy and accommodation costs, which is determined by the contract with the institute and the group rehabilitation organizers. In this programme, the institute determines the number of days for group rehabilitation and the funds to implement it in a specific year. In this case, it is again a very small group of physically handicapped seniors who are eligible for this type of care (ZZZS, 2018).

5.3.2 Mobility/accessibility adjustments

The next set of support mechanisms relates to health-related adjustments to programmes – above all mobility – thus facilitating physical access to the destination, and to various forms of adapted programmes. Slovenian seniors often state that transport to a holiday destination, or even to the place of departure for a vacation, becomes an obstacle. Travel agencies, for whom seniors represent an important target group, have made adjustments by offering vacations at the same time as their regular programmes – providing bus connections to all Slovenian cities from May to October, in three directions – from Murska Sobota, Carinthia and the Primorska regions. The purpose is to reach every major city, taking guests to their hotels and returning them (P.R. tourist agency representative, Ljubljana, 2016, personal communication). The Red Cross also organizes transport for socially deprived seniors. Transport to the holiday is organized in such a way that the bus route covers specific regions – for example, the Central Sava Valley and Gorenjska, Prekmurje and the Upper Soča Valley. Guests are brought almost directly to their accommodation (P.R., Žerjal, Ankaran, 2016, personal communication). They also emphasize the importance of transport in pensioners' associations, where they cooperate with specific travel agencies to organize daytrips or holidays – preferring to collaborate with those who offer a 'tailor-made' bus service (P.R. Škopelja, Škofja Loka, 2016, personal communication). Free door-to-door transport from home and back are also offered by individual Slovenian spas, such as Terme Dobrna, Terme 3000-Moravske Toplice, Radenci Health Resort, Terme Lendava, Terme Ptuj and Terme Banovci.

Programmes with other adjustments are also available. Adapting programmes for seniors from the perspective of content, scheduling and duration of visits and so on are of paramount importance (Carneiro *et al.*, 2013; Minnaert, 2014; Morgan *et al.*, 2015; Nikitina and Vorontsova, 2015). Commercial tourism providers, as well as all others, highlight various adjustments that contribute to the wider accessibility of tourist programmes, emphasizing, in particular, adjusting the pace of travel, sightseeing and selection of easily accessible attractions. Here, content adjustments based on the wishes of elderly guests (e.g. activities that promote socializing) are equally relevant and programmes can be fully adapted to the wishes of groups of seniors, such as the pensioners' associations.

The Premiki Institute and Travel Agency specifically offers extensive adaptation of tourist services for seniors, the physically handicapped and others. Their website details programmes that are adapted to groups (e.g. people who are physically challenged, have poor hearing, sight issues, etc.). Programmes for the elderly are adapted in terms of access to the tourist attractions, sufficient viewing time and programme content (P.R. Kores, Ljubljana, 2016, personal communication).

Even the holiday accommodation chosen by tourist agencies for the elderly should be tailored to their needs, in terms of easy access (elevators, ramps, without stairs) to accommodation, easy parking, proximity and easy access to important infrastructure, points of interest (e.g. easy access to the beach), adapted bathrooms (handrails) and so on.

5.3.3 Travel uncertainty adjustments

According to Minnaert (2014, p. 286), 'inexperience and a lack of knowledge can be linked to increased levels of uncertainty in making travel decisions'. As a result, it is very important to adapt the programmes for the broadest integration of the elderly into tourism. First, good information is required on the opportunities that older people have for becoming involved in such programmes. The adaptations revealed through our analysis provide a good description of the possibilities of integration in tourism thus reducing uncertainties (in this area). This includes, for example, detailed information about the programmes and their itineraries. The sindikalne-pocitnice.si web portal is a good example. The GIT trade union collates all union accommodation (union caravans, apartments and other holiday facilities) on a single site, representing affordable holiday offers for the elderly. The Seniorji.info web portal, a site dedicated to seniors, collates information on special offers, discounts and benefits in Slovenian spas, on a single website. It should not be forgotten, however, that it is also important to address seniors through channels that are not electronic, in the form of catalogues or brochures. Slovene pensioners' associations also play an important information-sharing role. Other aspects of reducing uncertainty, disclosed through interviews include, for example, 'all-inclusive' offers, where seniors need not worry about any of the aspects of the trip. Another good practice that has been revealed and could assist the elderly is to have an escort in addition to the tourist guide, such as a specific representative, for example, from a pensioners' society, who accompanies the group on holiday, and whom the members of the group know and trust.

5.4 Benefits

Contemporary literature points to the many benefits of social tourism for individuals, their quality of life and social inclusion (McCabe, 2009; Minnaert and Schapmans, 2009; Minnaert *et al.*, 2013; Gabruč, 2015). What do these kinds of programmes bring to seniors? At least three major groups of individual benefits have been identified through interviews with players who are directly or indirectly active in the field of social tourism for seniors in Slovenia.

Those interviewed identified the following benefits for participants taking part in such programmes, in particular: (i) social inclusion, or socializing (which is often highlighted as extremely important); (ii) broadening of horizons, knowledge – novel insights and experiences, involvement in various activities, quality leisure time; and (iii) maintaining health (physical and mental), and thus long-term active and healthy ageing, and subsequently maintaining autonomy and independence in old age. Tailor-made offers allow for greater choice or more options.

As has also been recognized in other related studies, tourism programmes offer opportunities for social interaction (Hazel, 2005; Minnaert *et al.*, 2009; McCabe *et al.*, 2010). Since loneliness and isolation can run in parallel with ageing, this kind of socializing is considered helpful (Morgan *et al.*, 2015). Taking part in social tourism programmes for the elderly encourages meeting new people, socializing and friendship. This aspect also proved to be important for the older social tourism users in our study, with the emphasis on the importance of conversations, group spirit, socializing, mutual encouragement and the positive motivation shared between them. In the framework of tourist agencies that develop targeted programmes for seniors, such programmes are regularly adjusted to include many social activities such as daytime and evening gatherings and group activities, through which close personal ties can develop. Additionally, holidays provided through pensioners' associations are organized in this way:

> The benefits? I think that that [a tourism programme] is absolutely necessary. People must also have time for themselves, to take care of themselves, and to meet their peers, to socialise and exchange opinions about small and ordinary daily matters and conversations. Different friendships are also made in this way. (spokesperson for the Association of Pensioners Škofja Loka)

Among the positive effects of social tourism is a broadening of horizons – acquiring new knowledge, getting to know new people and gaining new experiences and new perspectives, all of which can help in attaining new skills, influencing opinions, attitudes and, last but not least, behaviour (e.g. increased self-confidence). 'Social tourism is not just about broadening geographical horizons, but also mental and psychological' (Minnaert and Schapmans, 2009, p. 60). The possibility of involving seniors in tourism represents added value in their lives.

> With broadening horizons, life can be even nicer... [...] To give you some time, not only do you have a vacation, but an active vacation. [...] Wandering around an island, visiting foreign destinations, trying things out. I think it's all enriching. And this is like a University of the Third Age for tourism. (representative of the Trade Union of Catering and Tourism)

Taking part in tourism, changing the physical and social environment or visiting new places can improve general well-being, and either directly, or indirectly, positively influence one's general health. Research on integrating the elderly with tourism and self-perceived health, well-being, as well as functional abilities, shows positive relationships between one and the other. In this context, integrating seniors in tourism can be an important aspect of social and health policy, which actually reduces public spending on social and health care (Ferri *et al.*, 2013; Ferrer *et al.*, 2016), since social tourism can act as a preventative measure. The programmes of the Red Cross are perceived as such:

> And the help we offer in the social tourism for seniors segment is the opportunity to go to the coast, to change one's environment, to relax and experience the programmes that the health resort specifically offers them. And this certainly provides a few weeks of positive benefit to their health. (Red Cross spokesperson).

Maintaining health and care is perceived as beneficial for pensioners' well-being, as well as for the whole of society:

> It is generally beneficial for wider society. As you mentioned health tourism, it certainly brings to mind some kind of health preservation, which is generally positive for the whole of society, because a vital, satisfied pensioner who socialises is certainly more useful. (spokesperson for the Semedela Pensioners' Association)

In addition, there are other benefits, primarily economic benefits for providers, such as booking capacities out of season, connecting various actors (e.g. pensioners' associations and local providers, organizations with each other), and for pensioners' societies that increase their membership numbers and contributions from membership fees from organizing excursions. The benefits for society are predominantly preventative and curative activities in the field of seniors' health and thus economically beneficial to the state – bringing savings to the health insurance budget. Promoting or subsidizing social tourism has long-term economic benefits for both businesses and the state.

5.5 Conclusion and Perspectives for Future Development of Social Tourism for Seniors in Slovenia

Social tourism programmes for the elderly in Slovenia are currently left to individual and unrelated initiatives of various commercial and non-commercial providers of senior tourism. From the outset, a deliberation on future development and systemic measures in the field of senior social tourism must certainly consider the benefits that the state wants from such initiatives. It thus becomes necessary to decide whether to move forward on providing benefits primarily for vulnerable groups of seniors (the socially deprived, those with health problems [including mental health] and different types of disabilities). Incentives can be aimed at inclusion in tourism consumption primarily for those seniors who, owing to low social status, cannot afford holidays and the group of seniors who, because of medical conditions or different types of disabilities, have difficulty travelling. They may require a number of adjustments to undertake travel, for example, companion-aide, adapted infrastructure and other support services – initiatives that can be included within the adaptation model.

The second type of initiative focuses on the benefits for the entire senior population (and also includes vulnerable groups of seniors) – incentives in the context of inclusive social tourism. In principle, this is more inclusive and, as a result of economies of scale, also generates a number of economic benefits for tourism providers and destinations (such as reducing the effects of seasonality and preserving jobs). Additionally, providers can understand such initiatives within the stimulus model of social tourism.

Experiences of good practice within European social tourism show that these initiatives, unlike the first group of initiatives that are targeted (only) at vulnerable groups of seniors, are, as a rule, financially and socially sustainable (European Commission, 2010b). Therefore, as a strategic orientation, we propose measures primarily in the context of inclusive and stimulating social tourism models that promote both types of benefits – social and economic. In this way, the social and financial sustainability of such programmes can be maintained in the long run. Following the example of foreign good practice in social tourism (European Commission, 2010b), the proposed measures in the field of social tourism development for seniors in Slovenia (see Table 5.3) are designed as

Table 5.3. Overview of the proposed systemic measures. (Based on MGRT, 2017.)

Measure	Effects of measures
1 Development and establishment of the national social tourism system for seniors	• Development and promotion of Slovenian tourism through the promotion of social tourism programmes • Development of social tourism products • Meeting the needs of beneficiaries of social tourism and increasing accessibility to tourism • Reducing the effects of seasonality on the business of tourism companies • Increasing the quality of life and well-being of seniors
1.1 Preparation of a social tourism strategy for seniors	
1.2 Identification of key funding mechanisms for programmes	
1.2.1 Providing the necessary financial resources for (co-)financing social tourism programmes	
1.3 Identification and creation of key support mechanisms for the development of a national system of social tourism programmes	
1.3.1 Identification of key entities	
1.3.2 Designing social tourism programmes for seniors and defining beneficiaries together with entry conditions	
1.3.3 Establishing a network of involved commercial programme providers	
1.3.4 Designing a distribution network	
1.3.5 Selection of a private/commercial tourist company whose task would be to organize alterations and cooperation with the main programme administrator	
1.3.6 Defining the main partner/guardian/provider of social tourism programmes for seniors	
2 Web portal	
2.1 Creating a web portal	
2.2 Defining careers and policies and a plan of action	

multi-levelled and are linked to the following financial and support mechanisms:

1. Identifying key entities:

- the state/ministries and municipalities as (key) funders of the programmes;
- (non-)government/public agency, e.g. agencies in the public interest, participating in the design of programmes and defining the conditions for access by beneficiaries and providers into a national system of social tourism programmes;
- commercial organizations/agencies that take care of the business/economic aspects in a system of social tourism programmes; and
- (local) partners in the distribution of programmes (e.g. a network of agencies), information and support for the elderly (e.g. the Network of Pensioners' Associations).

2. Creating social tourism programmes for seniors and defining the group of beneficiaries, together with access conditions, and in the context of public–public partnerships (e.g. Ministry of Labour, Family, SocialAffairs and Equal Opportunities [MDDSZ] and Ministry of Economic Development and Technology [MGRT] in partnership with ZDUS).
3. Providing the necessary financial resources for (co-)financing these programmes (competent ministries – adopting and understanding social tourism programmes for seniors as part of social or healthcare policy).
4. Establishing a network of (inclusive) commercial programme providers.
5. Creating a reservation network (existing networks of pensioners' associations or a network of travel agencies).
6. Selecting private/commercial tourism companies (through public tender) whose task would be to organize shifts and cooperation with the main programme administrator – in the framework of a public–private partnership.
7. Defining the main partner/guardian/provider of the programmes (e.g. ZDUS). Through its network of pensioners' societies at the local level, this would be responsible for informing seniors about programmes and support for the elderly in preparing for holidays, forming groups and,together with a selected commercial company, administering the applications.
8. Creating an online portal with the administrator of the portal, who, in addition to the national system of social tourism programmes, also includes a parallel offer of social tourism programmes financed by humanitarian providers of social tourism programmes, or significant and exclusive discounts from commercial providers.
9. Implementing social tourism programmes, especially in the low season, from the beginning of October to the end of April, primarily in areas that are highly exposed to seasonal fluctuations (maritime and outlying destinations).

Below are some suggestions of systemic measures to ensure the sustainable development of social tourism for senior citizens in the Republic of Slovenia at the policy level (see Table 5.4).

Nine actions are presented at the level of (five) policies – social, health, education, tax policies, and policies in the field of transport and tourism. The whole system of measures supports the development of social tourism products to meet the needs and increase the accessibility of senior tourism, leading to improved quality of life and well-being. Indirectly, however, individual policies provide the basis for dealing with seasonality issues in tourism, contributing to the promotion of Slovenian tourism and increasing demand for Slovenian tourism products.

Table 5.4. Proposals for systemic measures at policy level. (Based on MGRT, 2017.)

Policy	Measure
Social policy	Financing of social tourism programmes as a measure of active employment policy. Reducing unemployment – the measure contributes to preserving old services and generating new, reducing the adverse effects of seasonality and its impact on employment
	The development and financing of social tourism programmes with an emphasis on affordable tourism programmes – the measure brings many benefits for individuals (with special needs) and society, reduces social risk and contributes to increasing social inclusion for (isolated) seniors and improving their quality of life
Health policy	Development and financing of social programmes focusing on programmes that bring health benefits – working preventatively, the measure brings about physical and mental health benefits in seniors
	Financing rehabilitation programmes and placing them in the context of (national) social tourism – the measure brings benefits to health-disadvantaged individuals within the elderly population
Educational policy	Developing and financing social programmes that promote learning – the measure brings benefits in the field of active living for the elderly
	Training of tourism workers for work in the field of social tourism – the measure encourages the development of specific competences of tourism personnel, thereby increasing their satisfaction with work and competitiveness in the labour market
Transport and tourism policy	Development of social tourism programmes or a national system of social tourism programmes (for seniors) – the measure contributes to the development of tourism in general and its promotion
Tax policy	Development of tax incentives for the subjects involved in the national system/social tourism programme (for seniors) – the measure encourages the attractiveness of seniors' tourism as a market-intensive business activity
	Development of tax incentives for included/potential funders of social tourism programmes – the measure encourages the attractiveness of seniors tourism as a market-intensive business activity

References

Bizjak, B., Knežević, M. and Cvetrežnik, S. (2011) Attitude change towards guests with disabilities: reflections from tourism students. *Annals of tourism research* 38(3), 842–857. DOI: 10.1016/j.annals.2010.11.017.

Carneiro, M.J., Eusébio, C., Kastenholz, E. and Alvelos, H. (2013) Motivations to participate in social tourism programmes: a segmentation analysis of the senior market. *Anatolia* 24(3), 352–366. DOI: 10.1080/13032917.2013.767212.

Darcy, S. (2010) Inherent complexity: disability, accessible tourism and accommodation information preferences. *Tourism Management* 31(6), 816–826. DOI: 10.1016/j.tourman.2009.08.010.

Diekmann, A. and McCabe, S. (2011) Systems of social tourism in the European Union: a comparative study. *Current Issues in Tourism* 14, 417–430.

Diekmann, A. and McCabe, S. (2013) Systems of social tourism in the European Union: a critical review. In: Minnaert, L., Maitland, R. and Miller, G. (eds) *Social Tourism: Perspectives and Potential.* Routledge, London, pp. 19–32.

Diekmann, A., McCabe, S. and Minnaert, L. (2012) Social tourism today: stakeholders and supply and demand factors. In: McCabe, S., Minnaert, L. and Diekmann, A. (eds) *Social Tourism in Europe: Theory and Practice.* Channel View Publications, Bristol, UK, pp. 35–48.

Domínguez Vila, T., Darcy, S. and Alén González, E. (2015) Competing for the disability tourism market – a comparative exploration of the factors of accessible tourism competitiveness in Spain and Australia. *Tourism Management* 47, 261–272. DOI: 10.1016/j.tourman.2014.10.008.

European Commission (2010a) Calypso study – compendium of good practices. Available at: https://ec.europa.eu/docsroom/documents/6925/attachments/1/translations/en/renditions/pdf (accessed 5 July 2019).

European Commission (2010b) Calypso study – final report. Available at: www.keroul.qc.ca/DATA/PRATIQUEDOCUMENT/29_fr.pdf (accessed 5 July 2019).

EUROSTAT (2018) Statistics explained. Population structure and ageing. Available at: https://ec.europa.eu/eurostat/statisticsexplained/index.php/Population_structure_and_ageing (accessed 9 October 2018).

Ferrer, J.G., Sanz, M.F., Ferrandis, E.D., McCabe, S. and García, J.S. (2016) Social tourism and healthy ageing. *International Journal of Tourism Research* 18(4), 297–307. DOI: 10.1002/jtr.2048.

Ferri, M., Durá, E. and Garcés, J. (2013) Functional health benefits for elderly people related to social tourism policy promotion. *International Journal of Multidisciplinary Social Sciences Issue* 1, 1–8.

Gabruč, J. (2014a) Socialni turizem v Sloveniji. [Social tourism in Slovenia.]. In: Vodeb, K. (ed.) *Trajnostni razvoj turističnih destinacij alpsko-jadranskega prostora. Založba Univerze na Primorskem*. Koper, Slovenia, pp. 83–103.

Gabruč, J. (2014b) Socialni turizem v Sloveniji. In: Mušič, K., Kociper, T. and Sikošek, M. (eds) *Turizem in management: na poti k uspešni poslovni prihodnosti. Založba Univerze na Primorskem*. Koper, Slovenia, pp. 113–125.

Gabruč, J. (2015) Socialni turizem in soustvarjanje socialnih koristi. [Social tourism and social benefits co-creation.]. In: Nemec Rudež, H. (ed.) *Soustvarjanje v turizmu. Založba Univerze na Primorskem*. Koper, Slovenia, pp. 44–63.

Hazel, N. (2005) Holidays for children and families in need: an exploration of the research and policy context for social tourism in the UK. *Children & Society* 19(3), 225–236. DOI: 10.1002/chi.838.

Leban, K. (2009) Družbena vloga in delovanje sindikatov na področju dopustovanja. Available at: www.arhiv.ds-rs.si/kb/posveti/upload/File/2009-09-21_Leban.doc (accessed 6 July 2019).

Lee, B.K., Agarwal, S. and Kim, H.J. (2012) Influences of travel constraints on the people with disabilities' intention to travel: An application of Seligman's helplessness theory. *Tourism Management* 33(3), 569–579. DOI: 10.1016/j.tourman.2011.06.011.

McCabe, S. (2009) Who needs a holiday? Evaluating social tourism. *Annals of Tourism Research* 36(4), 667–688. DOI: 10.1016/j.annals.2009.06.005.

McCabe, S., Joldersma, T. and Li, C. (2010) Understanding the benefits of social tourism: linking participation to subjective well-being and quality of life. *International Journal of Tourism Research* 12(6), 761–773. DOI: 10.1002/jtr.791.

McCabe, S., Minnaert, L. and Diekmann, A. (2012) *Social Tourism in Europe: Theory and Practice*. Channel View Publications, Bristol, UK.

MGRT – Ministrstvo za gospodarski razvoj in tehnologijo (2017) Identifikacija, klasifikacija in potenciali razvoja produktov socialnega turizma v Sloveniji, s poudarkom na programih za seniorje – Končno poročilo [Identification, classification and potential for the development of social tourism products in Slovenia, with an emphasis on programmes for senior citizens – final report.]. Available at: www.mgrt.gov.si/fileadmin/mgrt.gov.si/pageuploads/turizem/CRP/POROCILO_SKUPNO__dopolnitev_7112017__1___CRP_TURISTICA.docx (accessed 3 July 2019).

Minnaert, L. (2014) Social tourism participation: The role of tourism inexperience and uncertainty. *Tourism Management* 40, 282–289. DOI: 10.1016/j.tourman.2013.07.002.

Minnaert, L. and Schapmans, M. (2009) Tourism as a form of social intervention: the holiday participation centre in Flanders. *Journal of Social Intervention: Theory and Practice* 18(3), 42–61. DOI: 10.18352/jsi.171.

Minnaert, L., Maitland, R. and Miller, G. (2009) Tourism and social policy: the value of social tourism. *Annals of Tourism Research* 36, 316–334.

Minnaert, L., Maitland, R. and Miller, G. (2013) What is social tourism? In: Minnaert, L., Maitland, R. and Miller, G. (eds) *Social Tourism: Perspectives and Potential*. Routledge, London, United Kingdom, pp. 5–17.

Morgan, N., Pritchard, A. and Sedgley, D. (2015) Social tourism and well-being in later life. *Annals of Tourism Research* 52, 1–15. DOI: 10.1016/j.annals.2015.02.015.

Nikitina, O. and Vorontsova, G. (2015) Aging population and tourism: Socially determined model of consumer behavior in the 'senior tourism' segment. *Procedia - Social and Behavioral Sciences* 214, 845–851. DOI: 10.1016/j.sbspro.2015.11.736.

Pagán, R. (2012) Time allocation in tourism for people with disabilities. *Annals of Tourism Research* 39(3), 1514–1537. DOI: 10.1016/j.annals.2012.04.005.

Sindikalni počitniški domovi so spet zanimivi [Trade union holiday homes are interesting again] (2017) Radio Koper. Available at: https://www.rtvslo.si/radio-koper/prispevki/novice/sindikalni-pocitniski-domovi-so-spet-zanimivi/415773 (accessed 14 October 2020).

Struc, M., Zega, P., Geiger Bukovec, S., Perko, M. and Sánchez, K.E. (eds) (2013) *60 let za dobro otrok: 1953 – 2013. [60 years for the good of children.]*. Zveza prijateljev mladine Slovenije, Ljubljana, Slovenia.

SURS (2018) Mednarodni dan starejših. [International day of elderly.]. Available at: https://www.stat.si/StatWeb/News/Index/6952 (accessed 9 October 2018).

Župančić, K.K. and Puljić, M. (2017) Turizam u Jugoslaviji u 70ih. [Tourism in Yugoslavia in the 70s. *Rostra: časopis studenata povijesti Sveučilišta u Zadru* 8, 233–246.

ZZZS (2018) Obnovitvena rehabilitacija invalidov, udeležba v organiziranih skupinah za usposabljanje ter letovanje otrok. [Rehabilitation of the disabled, participation in organized groups for training and holidays for children.]. Available at: https://zavarovanec.zzzs.si/wps/portal/portali/azos/pravice_zdravstvenih_storitev/pravice_rehabilitacija (accessed 17 October 2020).

6 Social Tourism for Seniors in Spain: An Example to be Followed?

Elisa Alén González[1]*, Trinidad Domínguez Vila[1] and Nieves Losada Sánchez[2]
[1]University of Vigo, Vigo, Spain; [2]University of Trás-os-Montes e Alto Douro, Vila Real, Portugal

6.1 Introduction

One of the most important social transformations of this century is the ageing of the global population. The number of people aged over 60 is expected to double by 2050 and to triple by 2100, increasing from 962 million in 2017 to 2.1 billion in 2050 and 3.1 billion in 2100 (United Nations, 2017). The current over-60 population represents 13% of the total, with an annual growth of 3% (United Nations, 2017). Europe is the continent with the largest number of people in this segment, as they account for a quarter of the total population. Spain will be the oldest country in the world by 2050. It is estimated that the population over 65 in Spain will increase from 25% in 2029 to 38.7% in 2064 (Instituto Nacional de Estadística, 2014), although other organizations, including the United Nations, have another point of view and have calculated that 40% of the Spanish population will be aged over 60 by 2050 (United Nations, 2015).

Several factors have contributed to these trends, notably the increase in life expectancy and improvement in survival, which have led to increased longevity becoming widespread worldwide. These trends, together with the reduction in fertility and the phenomenon of global migration, have caused a change in demographic structures (United Nations, 2018). A wide range of actions and plans have been implemented so far in order to try to minimize the repercussions of this transformation. These have included the 1982 Vienna International Plan of Action on Ageing and the European Union's Horizon 2020 research and innovation plan, and the Department of Economic and Social Affairs (DESA) United Nations Programme on Ageing. All of them have clearly emphasized the importance of the elderly, the promotion of health and well-being in old age, and the protection of an enabling and supportive environment for the senior population.

The need for an active and healthy ageing that improves the quality of life of the elderly has been identified as a key element in connection with the trends described above. Costanza *et al.* (2007) noted that a feeling of well-being depends on the type and number of opportunities available to satisfy various human needs. In this way, quality of life is linked to the perceived satisfaction of an individual in relation to the surrounding life circumstances. Human needs can be largely grouped into four basic blocks: physiological needs, security, sense of belonging and self-esteem. The enjoyment of free time and leisure is therefore presented as a way of seeking greater satisfaction (Rubenstein, 1980), based

*Corresponding author: alen@uvigo.es

 Social Tourism: Global Challenges and Approaches (eds J. Lima and C. Eusébio)
DOI: 10.1079/9781789241211.0007

on the positive relationship that exists between overall satisfaction with life or well-being and the travel experience (Neal *et al.*, 2007), with positive effects such as improved health, and enhanced self-esteem and social capital.

Tourism generally influences the physical, mental and social well-being of elderly people (Vellas, 1986; Lee and Tideswell, 2005), as it impacts their subjective well-being, quality of life, self-perceived health and life satisfaction (Hunter-Jones and Blackburn, 2007; Dolnicar *et al.*, 2012; Hagger and Murray, 2013), as well as the creation and sharing of memories (Tung and Ritchie, 2011; Marschall, 2012). Given all this, and in light of the general lack of research on this specific topic (Sedgley *et al.*, 2018), the main objective of this study is to identify the determining factors and the main aspects to be developed to facilitate healthy and active ageing, thus improving the life quality of elderly people based on the Spanish government's social tourism programmes designed for this population group. This chapter starts with a contextualization of the social tourism concept, followed by a review of the public policies for active and healthy ageing and their impact on the tourism industry. Finally, an in-depth analysis of social tourism programmes in Spain is included.

6.2 Social Tourism

Article 24 of the Universal Declaration of Human Rights (1948) included the right to rest and leisure, and to a reasonable limitation of working hours and periodic paid holidays for all. Based on this right, social tourism focuses on guaranteeing that these opportunities are available to the most disadvantaged groups or enabling these groups to enjoy them. The European Economic and Social Committee (EESC) has acknowledged the great conceptual diversity surrounding social tourism, and proposed three requirements to be met by social tourism to be classified as such:

1. Real-life circumstances are such that it is totally or partially impossible to fully exercise the right to tourism. This may be due to economic conditions, physical or mental disability, personal or family isolation, reduced mobility, geographical difficulties, and a wide variety of causes which ultimately constitute a real obstacle.
2. Someone – be it a public or private institution, a company, a trade union, or simply an organised group of people – decides to take action to overcome or reduce the obstacle which prevents a person from exercising their right to tourism.
3. This action is effective and actually helps a group of people to participate in tourism in a manner which respects the values of sustainability, accessibility and solidarity (EESC, 2006).

That is, social tourism provides access to tourism for people who, owing to their personal or social conditions, encounter various obstacles or barriers to doing so, which has a number of positive social and economic effects.

The International Social Tourism Organisation (ISTO), an organization engaged in promoting social tourism, includes five criteria on which this type of tourism should be based: (i) the right to enjoy tourism for the greatest number of people; (ii) the contribution of tourism to socialization; (iii) the creation of sustainable tourism structures in the area involved; (iv) the contribution to employment and economic growth; and (v) the contribution to global development. However, although much progress has been made in this field, more is needed. There is a paucity of regulatory provisions on the matter, and a limited number of social tourism programmes available. However, studies and research on the topic are on the rise. One of the key factors which needs improving for the promotion of social tourism worldwide is to show the wide range of benefits it brings. These benefits are not only felt by the users themselves, but also by the various tourism destinations. These programmes help to offset tourism seasonality, create jobs,foster sustainable tourism, and showcase natural, social, cultural and heritage resources (EESC, 2006).

6.3 Policies for Active and Healthy Ageing: Their Impact on the Tourism Sector

Several studies have demonstrated the benefits of tourism on health and well-being (Richards,

1999; Gilbert and Abdullah, 2004; McCabe, 2009; Minnaert *et al.*, 2009; Dolnicar *et al.*, 2012; Lehto *et al.*, 2012). The use of holidays to enjoy free time and leisure provides both physical and mental benefits, as it improves people's quality of life, reduces stress, and encourages a more active and healthier life (Garcés *et al.*, 2016). As a general rule, after a pleasure trip, people feel healthier, happier and more relaxed (de Bloom *et al.*, 2011). Nevertheless, the above statements need to be qualified, given the lack of research on the relationship between people's health, lifestyles and travel experience (Chen and Petrick, 2013). This is particularly true of senior travellers, as existing research on this segment has mainly focused on profiles, behavioural habits and market characteristics (Alén *et al.*, 2017).

Social tourism, therefore, is a key element in social and health programmes, mainly for the elderly, owing to the continuous ageing of the population worldwide (Minnaert *et al.*, 2009). Senior tourists greatly benefit from these schemes, while also generating economic and social profitability for the destination, and they are used as a guide for the design and formulation of government policies (Backer and King, 2016). This impact is noticeable with respect to improved quality of life, active ageing, economic and social well-being of families, and efficiency of the state's social and health systems (Ferri *et al.*, 2013). That is, the individual feels that his life in the general environment is going well (Lucas *et al.*, 2004). More and more, it is argued that our feelings of well-being depend on the type and number of opportunities to meet various human needs (Costanza *et al.*, 2007). In this way, quality of life is related to the understanding of the perceived satisfaction of a person in relation to the circumstances in which he lives. Therefore, by reviewing the literature, it can be verified that there is a certain unanimity in considering four basic components of the quality of life: basic physiological needs (food and water, good health and physical protection against damage), security (a stable place to live and work, and opportunities to work and progress), belonging (links to social media networks and opportunities to participate in social, cultural and political activities) and self-esteem (knowledge and confidence, as well as the ability and freedom to decide).

For this reason, it is considered appropriate to examine the phenomenon of tourism as a leisure and experiential activity. In the developed world, even in cases where there seems to be high quality of life, time and resources are invested in abandoning the daily travelling environment. Leisure time, in general, and vacation, in particular, are increasingly being recognized as one of the experiences that people value in terms of their quality of life (Cushman *et al.*, 1996). Consequently, a lack of travel can lead to a decrease in quality of life (Neulinger, 1982). In summary, tourism can contribute to the quality of life of tourists, increasing their health by reducing tension and increasing physical activity, especially in areas without pollution; through the improvement of the self-esteem through a greater knowledge, competence and confidence; and also with the increase in social capital through the development of social networks and a better appreciation of the importance of participation in social, cultural and political activities.

While there are few studies in this field, and even fewer empirical ones, Ovsenik (2015), Tomka *et al.* (2015) and Woo *et al.* (2016) established three broad areas to be assessed by seniors in terms of how tourism can improve their quality of life through active and healthy ageing. The first of these is based on physical well-being, which focuses on the type, comfort and quality of the services provided, characteristics of accommodation and its surroundings. The second block concerns mental/emotional well-being, including personal development, independence, autonomy, learning and trust, among others. The third block, social welfare, focuses on social interaction, developing new relationships, participating in social, cultural and leisure activities and co-existence, among others.

Some examples of social tourism programmes have their origin in Europe, since social tourism is more widespread in the European Union (EU) countries (Backer and King, 2016), although these schemes also exist in Japan and North America (Fred, 2008; Seonhee, 2008). Of note are the 'IMSERSO Tourism Programme' in Spain; the 'INATEL Senior Health and Spa Programme' in Portugal; and the Calypso platform, which manages different actions and tourism programmes between different EU countries for young people, families with difficulties, elderly people and people with disabilities. Currently, these programmes are becoming obsolete, as the new generations of senior tourists, the so-called 'young seniors', are less interested in this type of travel (Chen and Shoemaker, 2014). The senior tourism segment has ceased to be a homogeneous group in its consumption patterns (Ryan and Trauer, 2005; van den Berg *et al.*, 2011; Le Serre and Chevalier, 2012). As they have greater purchasing power, they demand better quality and comfort in tourism products, and travel independently, to the detriment of tour packages (Alén *et al.*, 2017).

6.4 Social Tourism in Spain

6.4.1 The IMSERSO tourism programme

Spain is one of the leading countries worldwide in terms of social tourism programmes designed for seniors. The Institute for the Elderly and Social Services (IMSERSO) has used the various Programmes for the Promotion of Active Ageing to foster healthy retirement and ageing since 1985. These social tourism programmes seek to improve the quality of life of the elderly through tourism, promote employment in the sector, help businesses offset seasonality and boost economic activity in various sectors (Ministry of Health, Social Services and Equality, 2017).

According to the latest available data, two types of programmes were offered. One of them focused on social tourism and the other provided social spa treatments, which could be used by pensioners aged 55 or over, recipients of benefits or unemployment benefits aged 60 or over and holders or beneficiaries of the Social Security system, provided that they were 65 or older. They could be accompanied by their children (if they have a degree of disability equal to or greater than 45%), or by their spouse or significant other, even if they did not meet the age requirements (Ministry of Health, Social Services and Equality, 2017).

6.4.2 The 'social tourism programme'

The social tourism programme takes place between October and June each year, with the aim of combating the seasonality of tourism demand in Spain. In general, the packages offered can be divided into the following:

- *Holidays with or without transport* – 8-, 10- or 15-day full-board accommodation in seaside hotels either in mainland Spain or on the Spanish islands.
- *Cultural tours* – 5-, 6- or 7-day, full-board accommodation at different cultural destinations.
- *Nature tourism* – 4- or 5-day stays in natural settings, focused on recreational and contemplative activities.
- *Trips to provincial capitals* – 3- or 4-day half-board accommodation, with no pre-arranged transport, focused on sightseeing and cultural and leisure activities.
- *Trips to the autonomous cities of Ceuta and Melilla* – 4-day full-board accommodation, focused on sightseeing and cultural and leisure activities.

In general, the main seaside destinations in these programmes are located in Catalonia, Valencia, Murcia, Andalusia, the Canary Islands and the Balearic Islands, while inland tourism covers the whole of Spain. In the 2015/16 season, a total of 1,132,387 places were offered, 60% of them for Andalusia, the Balearic Islands and the Valencia region. The least popular choices were nature tourism (12,197 places available), provincial capitals (10,000 places available), and the autonomous cities of Ceuta and Melilla (1,100 places available). The programme funded approximately 20.87% of the total cost, which meant that each user, depending on the package selected, made an average contribution of between €141.77 and €585.90.

Almost 50% of the participants for that season were aged between 65 and 74 years (23.83% aged from 65 to 69 and 24.59% aged from 70 to 74), followed by those who were aged 75 to 79 (18.29%), and 80 to 84 (13.02%). There were more women (56.3%) than men. Their average monthly income was equal to or lower than the minimum retirement pension, that is, up to €742 (22.43%). The preferences according to destination area were: Andalusia (240,819 places occupied), Valencian Community (199,180) and Catalonia (111,829), with the Balearic and the Canary Islands far behind (74,062 and 52,686, respectively). There was a clear tendency to choose beach destinations with warm weather at the time of the programme (from autumn to spring). The main means of transport used were aeroplane (63.76%) and bus (27.72%). The programme generated a total of 6,187,704 stays across 338 hotels, which led to a total of 11,678 direct jobs. When considering all activities in the tourism sector, a total of 72,574 new jobs were generated (IMSERSO, 2017).

The programme was evaluated highly by users, as more than 97% rated it as good (57.33%) or very good (40%). The same applied to accommodation, as 94% rated it highly, and the vast majority (68.66%) gave a very good evaluation. Transportation was rated reasonably well, (85.32% of the respondents), although 54.66% rated it as being only 'good" (IMSERSO, 2017).

6.4.3 Social spa programme

Following the data from the last IMSERSO report (IMSERSO, 2017), the social spa programme was started with the aim of enabling elderly people to access specialized spa establishments to receive preventive, healing and rehabilitative treatment that would improve both their quality of life and the economic development of the destinations involved. The 2016 programme lasted from February until December, and offered 196,494 places in 98 spas, with a total of 1,936,834 stays. The duration of the stays ranged from 10–12 days, at an average cost to the user of €335.88, depending on the spa and duration. The places offered were mainly concentrated in Galicia (20%) and Catalonia (13.33%), the destinations with the greatest number of spas.

The programme offered accommodation on a full-board basis, and included leisure and free-time activities, an insurance policy and spa treatments. The spa treatment consisted of an initial medical examination, medical treatment and follow-up, and a final report. Bone and joint treatments were in the highest demand, as these were requested by more than 89% of the users. Other available treatments were aimed to alleviate respiratory, digestive, renal or urinary, dermatological and nervous conditions.

There were 277,800 requests, of which more than 55% were made by couples. A total of 59.17% of applicants were women and most were between 65 and 74 years old, followed by those aged between 75 and 79. The most significant difference between this group of applicants and those requesting to participate in the social tourism programme was their income level. The most highly represented were those who had incomes in the range €900.01–1,050 (12.05%), €1,050.01–1,200 (9.21%) and €12,00.01–1,350 (7.17%), although the trend continued that the largest proportion of applicants had incomes that did not exceed the minimum contributory retirement pension (27.8%). The occupancy rate of the programme was 93.68% of the places offered (IMSERSO, 2017).

A total of 179 complaints were received, mainly about accommodation (63) and catering (43), although there was a good level of satisfaction from almost 90% of the users, who were particularly pleased about the care given by the spas and the medical attention received, which they emphasized as being key elements of the treatment (IMSERSO, 2017). The complaints were concerned with the fact that many users could not access all the treatments that the spas offered free of charge, and also mentioned that the accommodation in some establishments was in different facilities from those where the treatment was provided. There were different opinions about their stays depending on the dates and the spa that the users had enjoyed. In general, food services received the lowest ratings in terms of satisfaction, although they had reasonably good ratings. Of interest was the fact that 44.36% of the programme users did not participate in the complementary leisure offering, despite the broad range of activities provided, including dancing, parties, day trips, gymnastics, bingo, board games, crafts and the like (IMSERSO, 2017).

6.4.4 Social tourism vs. independent travel

The senior tourism segment is becoming increasingly diverse, so there are a growing number of variables that affect their behaviour. Variables such as age, work status, structure and composition of the household, leisure activities at the destination, and pull and push factors are very heterogeneous, which means that establishing closed, specific products for this population group is increasingly difficult. Even allowing for this diversity, it can be stated that Spanish senior tourists with low income tend to opt for IMSERSO trips that seek active and healthy ageing, while those who travel independently have higher income, devote more of their time to physical activities and sports, as well as culture, nature and leisure, based on the results of the study by Alén *et al.* (2017). Older seniors usually choose tour packages as opposed to independent travel, and retired seniors clearly prefer the packages offered by IMSERSO. The elderly who live in single-person households, however, prefer pre-arranged travel, different from that offered by IMSERSO. Those who opted for the IMSERSO programmes spent more time on shopping, although those who travelled independently or through non-governmental packages sought higher quality and therefore more expensive products. Independent travel was chosen by those whose reason to travel was to visit friends and family or engaging in cultural, nature and leisure activities. Those who opted for package tours spent more time on pre-arranged trips, as opposed to independent travellers who, among others, focused more on physical activities.

6.5 The Impact of Social Tourism on Seniors' Quality of Life

Given the scarcity of empirical research on the impact that social tourism has on seniors' quality of life, a small exploratory analysis was performed on the variables that may influence the quality of life of elderly tourists when they participate in social spa programmes. A survey was conducted among 43 people aged over 55 years old who had taken some form of spa treatment in 2018. The data were collected between October and November 2018 through a survey conducted in the spas using convenience sampling. This non-probability sampling technique was used because of the difficulty of knowing the dates when the users of the programme booked into the spas. The main sociodemographic characteristics of the sample are shown in Table 6.1. It can be seen that the profiles identified were similar to those mentioned above in connection with the IMSERSO programmes.

The type of spa programme that the participants tended to choose was treatment and full-board accommodation (58.1%), as more than 75% of the respondents purchased these products through public or subsidized programmes. Some 55.8% were first-time users of this kind of product, while 34.9% had already participated in some kind of subsidized social tourism programme two to four times. Relaxation treatments were the most requested (44.2%), along with health treatments (30.2%), and both together (25.6%), with average durations ranging from 6 days (32.6%) to 12 days (41.9%). What 23.3% of the respondents valued most highly about the spa product was the combination of the treatment, accommodation and spa location, followed by treatment and accommodation (18.6% valued these the most highly), and treatment, accommodation and the company (11.6% valued these the most highly). Therefore, it can be concluded that treatment and accommodation were the most valued components, while 44.2% rated complementary leisure activities as the lowest item. Consequently, the behaviour patterns discussed in the previous points were repeated.

How spa tourism products influenced the quality of life of seniors was then analysed. Based on the theoretical review by Ovsenik (2015), Tomka *et al.* (2015) and Woo *et al.* (2016), the variables to be analysed were distributed into the following three blocks: physical, mental/emotional and social well-being. These variables were measured using a 7-point Likert-type scale, where 1 was 'Not important' and 7 was 'Very important'. To analyse the impact that these variables had on the seniors' quality of life, a global question was asked in which the respondents rated (on a scale from 1 to 10) how the enjoyment of the spa package had improved their quality of life. The data analysis method is based on an analysis of variance (ANOVA). This is used to determine whether there are any statistically significant differences between the means of two or more independent groups (Molugaran and Rao, 2017). The objective of the research was to understand whether the quality of life of seniors after a spa package differed based on physical, mental/emotional and social well-being. The factors with a *P*-value

Table 6.1. Sociodemographic characteristics.

Sociodemographic profile	% of the sample
Gender	
Male	41.9%
Female	58.1%
Age	
Aged between 55 and 65	46.5%
Aged over 65	53.5%
Level of education	
Primary school	23.3%
Secondary school	60.5%
University	16.3%
Employment status	
Retired	62.8%
Not employed	25.6%
Employed	11.6%
Marital status	
Single	9.3%
Married with children	7%
Married without children	81.4%
Widow/widower	2.3%
Monthly income	
Less than €600	4.7%
€600–1000	27.9%
€1001–1500	46.5%
€1501–2000	14%
More than €2001	7%
Dependency	
Yes	30.2% (Low level 25.6% and Medium 4.7%)
Not dependent	69.8%
Average trips per year	
One	39.5%
Between one and three	60.5%
More than three	0%

$P \leq 0.05$ based on the result of the ANOVA analysis were selected (Figueiredo Filho *et al.*, 2013). As already mentioned and shown in Table 6.2, the factors analysed are based on the studies of Ovsenik (2015), Tomka *et al.* (2015) and Woo *et al.* (2016).The use of ANOVA makes it possible to conclude whether the subjects who enjoyed the different spa programmes differ in the performance measure selected; that is, the three blocks that influence the quality of life of seniors: physical, mental/emotional and social well-being. The ANOVA is an omnibus test statistic and cannot tell you which specific groups were statistically significantly different from each other; it only tells you that at least two groups were different.

Table 6.2. Impact of physical, mental/emotional and social well-being on improving seniors' quality of life (ANOVA test).

Factors analysed	*F*	Significance
Physical well-being		
Comfort of the services provided	12.412	0.000
Quality of the services provided	8.194	0.000
Medical services available	2.356	0.071
Relaxation treatments	4.188	0.007
Medical treatments	4.852	0.003
Beauty treatments	1.351	0.269
Category of the establishment	1.351	0.029
Surroundings	0.788	0.758
Security in the establishment and its surroundings	4.854	0.003
Peacefulness of the establishment	3.318	0.020
Mental/emotional well-being		
Personal development	3.135	0.025
Independence	2.074	0.103
Autonomy	3.304	0.020
Free choice	4.873	0.003
Active learning	2.887	0.035
Confidence	5.041	0.002
Finding personal interests	2.499	0.059
Social well-being		
Social interaction with other people	1.020	0.410
Creating new bonds and relationships	1.704	0.169
Participating in social activities	1.409	0.250
Participating in cultural activities	2.076	0.103
Public spa programmes available	11.212	0.000
Staying with family	1.384	0.258
Staying with friends	1.343	0.272
Staying with strangers	1.888	0.133
Assessment of the enjoyment of leisure time	0.148	0.963

The results showed that the comfort and quality of the services offered, the relaxation and medical treatments, and the safety and peacefulness of the establishment and its surroundings were key factors for the improvement of life quality (Table 6.2). The individual characteristics of the specific establishment had no impact at all.

Regarding mental/emotional well-being, it was noted that almost all the analysed factors played a very important role, except for independence and finding personal interests. Confidence and freedom of choice, together with autonomy, personal development and active learning, were determining elements that had an impact on the mental and emotional well-being of seniors' quality of life. Finally, the analysis of the factors that constitute social well-being showed that they have little or no influence on seniors' enhanced quality of life, with the exception of public spa programmes, clearly linked to their therapeutic nature, rather than their social character (Table 6.2).

6.6 Conclusions

As the global population is ageing, it is necessary to improve the life quality of elderly people by developing active and healthy ageing policies that enhance the last stage of their lives. The enjoyment of free time and leisure, which is very important for this population group, should therefore be covered through public and private activities and programmes that are appropriate to their needs. Social tourism programmes are a decisive element in achieving these aims, as they provide physical, emotional/mental and social benefits for users, while they also benefit the economies of the managing countries.

Spain is one of the countries with the longest tradition of both national and international public social tourism programmes. By analysing the available options within these programmes, it can be seen that, despite having a wide segment of users with a diverse profile, the great majority demand a basic product (mainly accommodation, food services and environment/treatment according to the chosen programme) with high levels of quality and comfort. This is consistent with the results obtained in the exploratory analysis to determine the factors that influence seniors' quality of life, which also emphasized the importance of confidence and the individual's freedom of choice. It is striking that, while social well-being was a determining factor in the theoretical reviews on the subject, on a practical level it was not assessed by participants as being a key element in improving their quality of life.

Although there are several programmes with these characteristics, many of them have become obsolete or outdated, as they are focused on senior tourists who currently differ a lot from the senior tourists of a couple of decades ago. The senior population group is characterized by being a very heterogeneous segment, which increasingly results in a demand for products that are more suited to their needs and that provide them with new experiences. That is why governmental authorities must continue to work on the options provided within social tourism programmes for seniors, focusing on improving the quality and comfort of the product/basic package offered. The range of what is currently available also needs to be expanded to try to meet the demands of 'young seniors'. IMSERSO planned to meet this demand by offering trips abroad for Spanish seniors in 2019–2020, even though this would potentially be detrimental to the Spanish economy by not obtaining an economic return from some of the participants. In the event, the Covid-19 pandemic meant that all the trips were cancelled.

The study of seniors' quality of life through both publicly and privately managed tourist activities, and the determination of the current needs of senior tourists, must be explored further in future research. The limitations found are mainly down to the scarcity of studies to determine the factors that influence the quality of life of senior tourists, and the merely exploratory nature of the analysis. These limitations should be overcome by obtaining a larger sample.

References

Alén, E., Losada, N. and Domínguez, T. (2017) Activity of older tourists: Understanding their participation in social tourism programs. *Journal of Vacation Marketing* 23(4), 295–306.

Backer, E. and King, B. (2016) VFR traveller demographics: The social tourism dimension. *Journal of Vacation Marketing* 23(3), 191–204.

Chen, C.C. and Petrick, J. (2013) Health and wellness benefits of travel experiences: A literature review. *Journal of Travel Research* 52(6), 709–719.

Chen, S.C. and Shoemaker, S. (2014) Age and cohort effects: The American senior tourism market. *Annals of Tourism Research* 48, 58–75. DOI: 10.1016/j.annals.2014.05.007.

Costanza, R., Fisher, B., Ali, S., Beer, C., Bond, L. *et al.* (2007) Quality of life: An approach integrating opportunities, human needs, and subjective well-being. *Ecological Economics* 61(2-3), 267–276. DOI: 10.1016/j.ecolecon.2006.02.023.

Cushman, G., Veal, A.J. and Zuzanek, J. (1996) *World Leisure Participation: Free Time in the Global Village*. CAB International, Wallingford, UK.

de Bloom, J., Geurts, S.A.E., Sonnentag, S., Taris, T., de Weerth, C. *et al.* (2011) How does a vacation from work affect employee health and well-being? *Psychology & Health* 26(12), 1606–1622. DOI: 10.1080/08870446.2010.546860.

Dolnicar, S., Yanamandram, V. and Cliff, K. (2012) The contribution of vacations to quality of life. *Annals of Tourism Research* 39(1), 59–83. DOI: 10.1016/j.annals.2011.04.015.

EESC (2006) Opinion of European economic and social Committee on social tourism in Europe. *Official Journal of European Union C318/67*.

Ferri, M., Durá, E. and Garcés, J. (2013) Functional health benefits for elderly people related to social tourism policy promotion. *International Journal of Multidisciplinary Social Sciences* 1, 1–8.

Figueiredo Filho, D.B., Paranhos, R., Rocha, E.C., Batista, M., Silva Jr., J.A. *et al.* (2013) When is statistical significance not significant? *Brazilian Political Science Review* 7(1), 31–55. DOI: 10.1590/S1981-38212013000100002.

Fred, E. (2008) *Social Entrepreneurship and Urban Tourism as Economic Development: Best Practices from Long Beach*. University of Southern California, Los Angeles, California.

Garcés, J., Ferri, M., Durá, E., McCabe, S. and Sánchez, J. (2016) Social tourism and healthy ageing. *International Journal of Tourism Research* 18, 297–307.

Gilbert, D. and Abdullah, J. (2004) Holiday taking and the sense of well-being. *Annals of Tourism Research* 31(1), 103–121. DOI: 10.1016/j.annals.2003.06.001.

Hagger, C. and Murray, D. (2013) Anticipating a flourishing future with tourism experiences. In: Filep, S. and Pearce, P. (eds) *Tourist Experiences and Fulfilment: Insights from Positive Psychology*. Routledge, Abingdon, UK, pp. 186–202.

Hunter-Jones, P. and Blackburn, A. (2007) Understanding the relationship between holidaytaking and self-assessed health: An exploratory study of senior tourism. *International Journal of Consumer Studies* 31(5), 509–516. DOI: 10.1111/j.1470-6431.2007.00607.x.

IMSERSO (2017) *Informa Anual IMSERSO 2017. Programas de promoción del envejecimiento activo*. Ministerio de Sanidad, Servicios Sociales e Igualdad, Madrid.

Instituto Nacional de Estadística (INE) (2014) *Proyección de la población de España 2014–2064*. INE, Madrid.

Le Serre, D. and Chevalier, C. (2012) Marketing travel services to senior consumers. *Journal of Consumer Marketing* 29(4), 262–270. DOI: 10.1108/07363761211237335.

Lee, S.H. and Tideswell, C. (2005) Understanding attitudes towards leisure travel and the constraints faced by senior Koreans. *Journal of Vacations Marketing* 11(3), 249–263. DOI: 10.1177/1356766705055716.

Lehto, X.Y., Lin, Y.-C., Chen, Y. and Choi, S. (2012) Family vacation activities and family cohesion. *Journal of Travel & Tourism Marketing* 29(8), 835–850. DOI: 10.1080/10548408.2012.730950.

Lucas, R.E., Clark, A.E., Georgellis, Y. and Diener, E. (2004) Unemployment alters the set point for life satisfaction. *Psychological Science* 15(1), 8–13. DOI: 10.1111/j.0963-7214.2004.01501002.x.

Marschall, S. (2012) Tourism and memory. *Annals of Tourism Research* 39(4), 2216–2219. DOI: 10.1016/j.annals.2012.07.001.

McCabe, S. (2009) Who needs a holiday? Evaluating social tourism. *Annals of Tourism Research* 36(4), 667–688. DOI: 10.1016/j.annals.2009.06.005.

Ministry of Health, Social Services and Equality (2017) *Informe annual IMSERSO 2016*. Secretaria de Estado de Servicios Sociales e Igualdad, IMSERSO.

Minnaert, L., Maitland, R. and Miller, G. (2009) Tourism and social policy: The value of social tourism. *Annals of Tourism Research* 36(2), 316–334.

Molugaran, K. and Rao, G.S. (2017) *Statistical Techniques for Transportation Engineering*. Elsevier, Oxford.

Neal, J.D., Uysal, M. and Sirgy, M.J. (2007) The effect of tourism services on travelers' quality of life. *Journal of Travel Research* 46(2), 154–163. DOI: 10.1177/0047287507303977.

Neulinger, J. (1982) Leisure lack and the quality of life: The broadening scope of the leisure professional. *Leisure Studies* 1(1), 53–63. DOI: 10.1080/02614368200390051.

Ovsenik, R. (2015) Quality of life and senior travel motivations and expectations. *Mediterranean Journal of Social Sciences* 6(4), 590–599. DOI: 10.5901/mjss.2015.v6n4s2p590.

Richards, G. (1999) Vacations and the quality of life: Patterns and structures. *Journal of Business Research* 44, 189–198.

Rubenstein, C. (1980) Vacations: Expectations, satisfactions, frustrations, fantasies. *Psychology Today* 14, 62–76.

Ryan, C. and Trauer, B. (2005) Ageing populations: Trends and the emergence of the nomad tourist. In: Theobald, W.F. (ed.) *Global Tourism*, 3rd edn. Butterworth-Heinemann, Boston, Massachusetts, pp. 510–528.

Sedgley, D., Haven-Tang, C. and Espeso-Molinero, P. (2018) Social tourism & older people: The IMSERSO initiative. *Journal of Policy Research in Tourism, Leisure and Events* 10(3), 286–304. DOI: 10.1080/19407963.2018.1465064.

Seonhee, J. (2008) Assessing social capital and community involvement: Social network analysis for the sustainable Amish community tourism development. PhD thesis. University of Illinois at Urbana-Champaign, Urbana-Champaign, Illinois.

Tomka, D., Holodkov, V. and Andjelkovic, I. (2015) Quality of life as a travel motivational factors of senior tourists – results of research in Novi Sad. *Informatologia* 40(1–2), 62–70.

Tung, V.W.S. and Ritchie, J.R.B. (2011) Exploring the essence of memorable tourism experiences. *Annals of Tourism Research* 38(4), 1367–1386. DOI: 10.1016/j.annals.2011.03.009.

United Nations (2015) *World population ageing 2015. ST/ESA/SER.A/390, Department of Economic and Social Affairs, Population Division*. United Nations, New York.

United Nations (2017) *World population prospects. Key findings & advance tables. 2017 Revision. ESA/P/WP/248, Department of Economic and Social Affairs, Population Division*. United Nations, New York.

United Nations (2018) *Ageing. Global issues overview*. United Nations, New York.

van den Berg, P., Arentze, T. and Timmermans, H. (2011) Estimating social travel demand of senior citizens in the Netherlands. *Journal of Transport Geography* 19(2), 323–331. DOI: 10.1016/j.jtrangeo.2010.03.018.

Vellas, P. (1986) *Aspectos psicológicos y sociales del desarrollo del turismo en la tercera edad. Primeras Jornadas Técnicas de Turismo y Tercera Edad*. IMSERSO, Alicante.

Woo, E., Kim, H. and Uysal, M. (2016) A measure of quality of life in elderly tourists. *Applied Research Quality Life* 11(1), 65–82. DOI: 10.1007/s11482-014-9355-x.

7 '*We need a vacation*': Social Tourism Initiatives for Low-Income Families with Children with Disabilities

Celeste Eusébio[1]* and Mariana Pedrosa[2]
[1]*Department of Economics, Management, Industrial Engineering and Tourism, University of Aveiro and GOVCOPP Research Unit, Aveiro, Portugal;* [2]*Department of Economics, Management, Industrial Engineering and Tourism, University of Aveiro, Portugal*

7.1 Introduction

Although the United Nations (UN, 1948) declared leisure and holidays a human right, there are still many people excluded from tourism owing to various constraints, such as physical disabilities, poor economic conditions, reduced mobility and isolation (Eusébio *et al.*, 2013). Among the groups identified as disadvantaged regarding their access to tourism, seniors (Eusébio *et al.*, 2013), young and disabled people (Darcy *et al.*, 2010; Figueiredo *et al.*, 2012), and, more recently, low-income families (Minnaert *et al.*, 2009) and families whose children have disabilities (Kim and Lehto, 2013) have received some academic attention. However, studies about the tourism experiences of low-income families with children with disabilities (LIFWCD) and regarding the role of social tourism programmes (STP) as a facilitator to allow this market segment access to tourism activities are very scarce.

LIFWCD are doubly penalized regarding their access to tourism experiences. In addition to all the travel constraints of a disabled person, strong financial constraints also emerge as one of the most important constraints, contributing to excluding this market segment from tourism. However, empirical evidence reveals that participating in tourism promotes enrichment of a society in several domains (social, cultural, economic and political) (Eusébio *et al.*, 2013; Ferrer *et al.*, 2016). Consequently, promoting initiatives that overcome the barriers for LIFWCD to participate in tourism activities will contribute to promoting a more inclusive society.

The promotion of social policies that encompass tourism as a tool to stimulate the well-being of citizens and simultaneously to promote the economic and social development of the origin and destination countries has received some attention in some European and American countries (e.g. France, Brazil, Chile, Spain, Portugal and Belgium). However, there are differences in the level of support given to social tourism, with some countries actively boosting its development and others adopting a non-interventionist approach (McCabe, 2009; Minnaert *et al.*, 2011; Carneiro *et al.*, 2013). Moreover, international organizations (e.g. the International Social

*Corresponding author: celeste.eusebio@ua.pt

DOI: 10.1079/9781789241211.0008

Tourism Organisation [ISTO] and the European Union [EU]) have developed initiatives in this field (e.g. Calypso Project) that clearly show the relevance of social tourism initiatives (STI) in several domains (e.g. social, cultural, economic and political). Some STI have been promoted, mainly for seniors and young people in Portugal. However, some disadvantaged groups, such as LIFWCD, have been almost excluded from social policies that use tourism as a tool to promote inclusion. Therefore, it is important to design and implement instruments to stimulate the access to tourism activities for this group, converting their dream of obtaining tourism experiences into a reality.

Based on the above-mentioned arguments, it is urgent to carry out studies to provide knowledge regarding the tourism experiences of LIFWCD and to design suitable strategies to overcome the strong constraints that this group faces in assessing tourism activities. In this context, the aims of this study are: (i) to identify the travel constraints of LIFWCD; (ii) to characterize the tourism experience of these families; (iii) to identify the benefits obtained from family trips; and (iv) to identify guidelines to develop STP to promote the access of this market to tourism activities. To accomplish these aims, this chapter encompasses a literature review regarding STI, travel constraints and tourism experiences of LIFWCD and an empirical study. This empirical study was carried out with a group of Portuguese LIFWCD, specifically with parents, and with a group of social organizations that provide support to these families. The experience of the parents and their interpretation of their children's experiences were analysed, given their relevance in the travel decision-making process and during the trip.

7.2 Literature Review

Data provided by the World Bank and United Nations Educational, Scientific and Cultural Organization (UNESCO) point towards about 200 million children with disabilities worldwide or 10% of the world's youth population (Kim and Lehto, 2013). Families with disabled children face several challenges, one of these being a decrease in their income for two reasons: (i) the extra time necessary for care and thus less time available for parents to have paid work; and (ii) higher expenses to look after their children (Valentine, 2001). Consequently, people with disabilities constitute one of the most socially and economically excluded groups of our societies (Murray and Sproats, 1990; Fontes *et al.*, 2014). Although the analysis of disability as a social issue has received increased attention both in disability studies and in tourism research in recent years (e.g. Blichfeldt and Nicolaisen, 2011; Figueiredo *et al.*, 2012; Kim and Lehto, 2013; Kastenholz *et al.*, 2015; Nyman *et al.*, 2018), studies focusing on families with disabled children are very scarce.

The tourism experience of LIFWCD has been highly neglected in the tourism literature (Kim and Lehto, 2013), with only some research published in the field of leisure and recreation studies (e.g. Townsend *et al.*, 2010). Moreover, while research on family tourism has increased considerably in recent years, a limited number of studies has examined the tourism practices of families with children with disabilities (e.g. Hazel, 2005; Kim and Lehto, 2013; Nyman *et al.*, 2018). Hazel (2005) analyses the holidays for children and families in need. Kim and Lehto (2013) investigate the tourism motivations and activities of Korean families with disabled children. Nyman *et al.* (2018) carried out a qualitative study in Sweden (2018) to investigate the tourism travel constraints of families with wheelchair-bound children with cerebral palsy and how these travel constraints influence their destination choice.

The difficulty in gathering empirical data from the disabled population and their families may be one of the main reasons for the limited number of empirical studies published in this field (Kim and Lehto, 2013). However, the situation is even worse in Portugal (Fontes *et al.*, 2014), where systematic scientific studies on this topic are almost non-existent and largely restricted to the fields of psychology, medicine and education sciences. With the exception of some studies that have analysed the tourism experiences of a group of people with disabilities (e.g. Figueiredo *et al.*, 2012; Kastenholz *et al.*, 2015; Deville and Kastenholz, 2018; Moura *et al.*, 2018), the tourism literature in this field is very limited.

Despite the limited research regarding the tourism experience of LIFWCD, some studies recommend family leisure, recreation and tourism activities as relevant for overcoming the pressure resulting from disability and increasing the quality of life and well-being of children with disabilities and their families (Hazel, 2005; Townsend *et al.*, 2010; Figueiredo *et al.*, 2012; Kim and Lehto, 2013).

According to the family system theory, families are defined as interacting members, with each member influencing and being influenced by every other member of the system (Lima *et al.*, 2012; Lima, 2015). Therefore, a family is greater than the sum of its parts, where its healthy functioning is based on a balance between three dimensions: (i) family cohesion (togetherness or bonding); (ii) adaptability (ability to be flexible and cope with change); and (iii) family communication. Tourism experience may have an important role in all of these dimensions (Hazel, 2005; Townsend *et al.*, 2010; Eusébio and Carneiro, 2011, 2014). However, a great number of families with children with disabilities do not have the resources or conditions to participate in tourism activities. This situation is worse in the case of low-income families. As suggested by Kim and Lehto (2013, p. 14) 'differences in physical, mental, or emotional conditions between families with and without disabled children may lead different needs, interests, and limitations when it comes to participation in tourism activities'. Thus, it is of the highest relevance to examine the needs, motivations, constraints and benefits obtained by these families in terms of tourism experiences.

The family plays an important role in providing opportunities for personal development of children with disabilities (Figueiredo *et al.*, 2012; Kastenholz *et al.*, 2012; Kim and Lehto, 2013). The limited empirical studies published (e.g. Mactavish and Schleien, 2004; Townsend *et al.*, 2010; Kim and Lehto, 2013) show important tourism benefits for both children and their families, such as: (i) developing physical skills and competences; (ii) enhancing family bonds; (iii) promoting children's intellectual skills; (iv) improving creative and emotional development; and (v) helping family members develop various social skills. For example, studies carried out by Minnaert *et al.* (2009) and Smith and Hughes (1999) conclude that disadvantaged social groups such as economically disadvantaged families draw the same benefits from tourism activities as the general population, but this market feels the benefits more intensely (Smith and Hughes, 1999; Minnaert *et al.*, 2009; Lima *et al.*, 2011). Moreover, some studies also have demonstrated the important role of recreational and leisure activities in reducing stress and other difficulties resulting from the incapacity of the children (Glidden, 1993). The practice of tourism activities contributes to improving the quality of family relationships and consequently the quality of life of families. In the long term, it will lead to improvements in the physical and mental well-being of children and other household members (Larson and Verma, 1999). The learning opportunity created by tourism (Mitchell, 1998; Lima *et al.*, 2011), both for parents and for children with disabilities, is also an important benefit that these families can obtain through the practice of tourism activities.

Despite the great potential of tourism to improve the well-being and quality of life of parents and children with disabilities, and the desire of these families to participate in tourism activities, a great number of these families are excluded from tourism activities, mainly owing to several constraints. Although all people face travel constraints, some groups of people, such as families with young children with disabilities, face more than others (Nyman *et al.*, 2018). Therefore, for families with disabled children, tourism trips can require more negotiation strategies and adjustments than for other groups of visitors (Kim and Lehto, 2013; Nyman *et al.*, 2018). According to Smith (1987), travel constraints for people with disabilities can be categorized into three categories: intrinsic (related to visitors' psychological functioning and also economic limitations), interactive (resulting from the interaction with other people such as difficulty in communicating with others) and environmental (related to inaccessible tourism services, including inadequate attitudes by tourism services providers). Moreover, Nyman *et al.* (2018, p. 34) highlight, 'the disability of the child can be a constraint for the family but, on the other hand, the family can also be a constraint for the disabled child'. Therefore, the decision to carry out a specific tourism trip should be a 'trade-off between constraints and desires of the whole family' (Nyman *et al.*, 2018, p. 35).

The various difficulties faced on a daily basis by LIFWCD, mainly economic difficulties, lead them to situations of stress and attrition that sometimes cause them to have low self-esteem (Dowling and Dolan, 2010). Some studies published (e.g. Dobson and Middleton, 1998; Oldman and Beresford, 1998; Rickford, 2000) examine the effect of poverty on these families, highlighting the difficulty in paying for the therapies necessary for the growth of the children or participating in recreation and tourism activities (Dowling and Dolan, 2010).

In order to increase the participation of LIFWCD in tourism activities, it is of the greatest importance to promote STI, for example STP, for this market. Hazel (2005, p. 226) highlights this idea, mentioning that 'the aim of all parents and children having a regular family holiday has now widely been mainstreamed within the social care policies of many governments'. Usually this type of tourism is innovative, collaborative and may involve a large number of organizations (McCabe *et al.*, 2011). However, although there are social organizations (SOs) that have been providing support to disabled people in Portugal since the 1920s and 1930s (Fontes *et al.*, 2014), few organizations provide STI for disabled children and their families. Some studies (e.g. Fontes *et al.*, 2014) highlight that, owing to the precarious living conditions of the majority of disabled people in Portugal, social organizations (SOs) have focused their actions on improving living conditions, ignoring the relevance of providing tourism opportunities (e.g. Carneiro *et al.*, 2013; Eusébio and Carneiro, 2014). However, it is well recognized that tourism has an important role in the promotion of more inclusive societies. In this line of thought, Kim and Lehto (2013) remark that the differences in mental, physical or emotional conditions between families with and without disabled children probably give rise to different needs, interests and limitations regarding participation in tourism activities. Therefore, studies that address characteristics of this segment, their needs, types of tourism participation and benefits obtained, based on their real travel experience, are of utmost importance, not only for the tourism supply agents (private and public), but also for the agents responsible for defining social policies that promote the well-being and social inclusion of the disadvantaged groups of our society.

STI are one way of overcoming the various constraints that LIFWCD face in participating in tourism. The literature about the role of tourism as a form of social intervention (e.g. Higgins-Desbiolles, 2006; Minnaert and Shapmans, 2009; Diekmann and McCabe, 2011; Lima *et al.*, 2012; Eusébio *et al.*, 2013; Minnaert, 2014) highlights the benefits of STI for promoting more inclusive societies. However, the current economic climate has inspired debate in many countries about the state's role in providing social security and public welfare, frequently calling for private stakeholders to assume a more active role in resolving society's problems (Minnaert, 2014). Consequently, and in the face of many other more urgent social worries, any support of STI requires the evaluation of the real net benefits provided by these initiatives. The literature in this field is still scarce, but, in recent years, some European projects (e.g. Calypso) and publications (e.g. Minnaert *et al.*, 2009) have been developed in this area. There are a large variety of STI offered for disadvantaged groups worldwide. In several European countries (e.g. France, Belgium and Spain) social tourism is supported by public funding, while in other countries (e.g. the UK) STI are also supported by private agents, while in other countries (e.g. the USA) social tourism is less common. Portugal has some tradition in STI, with the INATEL Foundation playing a crucial role, offering several STP for disadvantaged groups. However, STI for LIFWCD are very limited.

7.3 Methodology

Socially excluded groups are likely to be excluded from holidays and their benefits. As seen in the literature review, social invisibility of LIFWCD is reproduced extensively in the social sciences, which have invested very limited efforts in analysing this economically and socially disadvantaged group. However, some evidence suggests that STI are urgently needed to overcome the constraints that this group faces in participating in tourism and to provide access to travel and leisure opportunities. In order to fill these gaps, this study intends to debate and clarify the relationships between the characteristics and challenges of a social group frequently excluded from tourism (LIFWCD), the travel constraints faced by this group, the travel behaviour of this group and the promotion of STI to enhance their access to tourism. To achieve this goal, an empirical study was conducted in Portugal to provide knowledge regarding the following topics: (i) constraints, motivations and travel behaviour of low-income families with disabled children; (ii) benefits obtained from participation in tourism activities; and (iii) guidelines to design STI to increase the participation of LIFWCD in tourism activities.

The methodology used in this research encompasses three stages. First, a search in journals, books, reports, papers and websites was carried out to identify social organizations that offer STI to this market in Portugal. Second, a social organization survey was undertaken via semi-structured interviews in order to identify and characterize the STI offered to this market and also to obtain the perceptions of these organizations regarding the relevance of these STI in improving the quality of life of these families. Third, a survey of Portuguese low-income families with disabled children was carried out, also through semi-structured interviews, to identify travel constraints, travel behaviour, benefits obtained from their participation in tourism activities and regarding the characteristics that STI should have to satisfy the needs of this segment.

There is no official register of the social organizations that provide support to low-income families with disabled children in Portugal. Consequently, an internet search was carried out and 47 Portuguese social organizations that support children with disabilities were identified to participate in this research. First, in November 2014, an email explaining the aims of this study and requesting the collaboration of each organization to participate in this survey was sent. However, only ten showed willingness to participate in this survey.

Given that there are no official data regarding the number of LIFWCD in Portugal, a snowballing sample approach was used to select a group of Portuguese low-income families to participate in a semi-structured interview. A total of 11 families with children with disabilities from different regions of Portugal (Vila Nova de Gaia, Porto and Mogadouro) were interviewed during the month of August of 2016.

The interview guide was designed to obtain information concerning: (i) initiatives offered by the social organizations to promote the participation of LIFWCD in tourism activities; (ii) level of participation in tourism activities; (iii) travel constraints; (iv) travel behaviours; (v) benefits obtained (for the family and for disabled children); (vi) interest and level of participation in STI; and (vii) sociodemographic characteristics of the family and features of the children's disabilities. Content analysis was used to analyse the data obtained in the interviews.

7.4 Analysis and Discussion of Results

7.4.1 Sample profile

Social organizations

The majority of social organizations that participated in this study work with children and youth (50%); only one of the social organizations works only with children. Regarding the types of users' disabilities, all organizations work with people with cognitive disability. Only two organizations work with all types of disabilities (Table 7.1).

LIFWCD interviewed

Almost all interviews were conducted with mothers, with only one interview being conducted with a father. Regarding the children's age, six had children between the ages of 16 and 29, with the remainder having children under 16 years of age. Concerning the disability of their children, the majority have cognitive disabilities. However, almost half of the sample (five families) have children with multideficiency (Table 7.2). This characteristic considerably influences the constraints that these families face when they carry out a tourism trip.

7.4.2 Needs, constraints and benefits of tourism activities for children and their families: The view of social organizations

The social organizations that were interviewed carry out several activities with disabled children. Sports activities (e.g. swimming and horse riding) are the most mentioned (60% of the institutions offer this type of activities). Cultural activities are offered by 40% of the institutions

Table 7.1. Characterization of the organizations interviewed.

Code	Type of organization	Age of the users	Type of users' disability
SO1	Private social solidarity institutions	Children and youth	Cognitive and mobility
SO2	Private social solidarity institutions	Children and youth	Mobility, cognitive and visual
SO3	Private social solidarity institutions	Children, youth and adults	Cerebral palsy
SO4	Private institution	Children, youth and adults	Cognitive
SO5	Private social solidarity institutions	Children, youth and adults	All types of disability
SO6	Private social solidarity institutions	Children	Cognitive and mobility
SO7	Private social solidarity institutions	Children and youth	Cognitive
SO8	Private social solidarity institutions	Children and youth	Cognitive
SO9	Private social solidarity institutions	Youth and adults	All types of disability
SO10	Private school	Children and youth	Autism

(e.g. music, going to the cinema, going to the theatre). Holiday camps are another type of tourism activity that about 30% of the institutions carry out. According to the opinion of those in charge of these social organizations, recreation and leisure activities are very important to the well-being and personal development of these children. The following quotes support this conclusion.

Table 7.2. Characterization of the families interviewed.

Code	Interviewee	Age of the child	Type of child's disability
F1	Mother	13	Mobility
F2	Father	12	Cognitive
F3	Mother	22	Mobility and visual
F4	Mother	23	Mobility and visual
F5	Mother	21	Mobility and cognitive
F6	Mother	9	Cognitive
F7	Mother	18	Cognitive
F8	Mother	25	Cognitive
F9	Mother	12	Hearing, mobility and cognitive
F10	Mother	21	Mobility
F11	Mother	15	Mobility and cognitive

> Leisure activities are very important, especially for the occupation of leisure time, in order to avoid the monotony of days and sedentariness. (I7)
>
> Recreational and leisure activities are one of the best vehicles for accessing new experiences and developing social and relational skills. (I2)

Another concern of these institutions is to help many children or young people to work and develop everyday skills on a personal, social or even school level. This concern is clear in the following quotes.

> Much of our work consists of an intensive intervention in the child's natural environment, coordinating the family, school and the community. (I4)
>
> The leisure activities we offer always have a therapeutic component and development of competences: (...) travel experience in a less usual means of transportation (...), moments in which the autonomy and activities of daily living are worked out, such as preparation of snacks or lunch. (I2)
>
> We have holiday camps available where recreational and sports activities are developed with the main objective of social inclusion. (I8)

The participation of the parents in the leisure and recreation activities is also mentioned by those in charge of the social organizations that we interviewed. Seven of the organizations facilitate the participation of the parents in the activities that they carry out, while three of them only permit this participation in cases where children need great support. The importance of the participation of parents in the activities carried out by children with disabilities is highlighted in the following quotes.

> The surfing involves parents initially to experience and perceive the richness of the activity and in some cases to help the children (...). But in general it is always positive and very enriching, both for the children and for their parents who have here the unique possibility of sharing the same type of recreation activity and also the same space. (I10)

> Parents participate in activities such as the ranch. There are also events that take place daily from 5:00 p.m. to 9:00 p.m. at weekends and a 'family space' where parents interact with the children, identifying the needs of each one. (I9)

One of the institutions that participated in this study also refers to the difficulty of involving the families in the activities developed, given that there are some cases of abandonment of children by their parents or relatives.

Although the social organizations consider as extremely important that children with disabilities have the same opportunities as children without disabilities, the number of leisure and recreation activities offered to this group is still very limited. Moreover, these families face several intrapersonal, interpersonal and structural constraints to participating in these activities and carrying out tourism trips.

7.4.3 Needs, constraints and benefits of tourism activities for children and their families: The view of the parents

Going to the beach and the swimming pool are the most common leisure and recreation activities carried out by the families interviewed, corroborating the results obtained in the study developed by Kim and Lehto (2013). Moreover, a great number of families reveal that their children's disability does not deprive them of holidays. However, it is not easy due to the lack of accessibility of tourism destinations. Unfortunately, this reality is not observed in all families interviewed. One of the families mentioned that they do not participate in leisure and recreation activities together with their child owing to the lack of physical accommodations at the destinations to receive children with disabilities. The difficulties that this family faces when they carry out a tourism trip are made explicit in the following quote.

> In a tourism destination, the tourism services providers are not prepared to offer activities for people with cerebral palsy. There is no equipment available: lack of ramps to facilitate access, unqualified staff, restaurants are not equipped with the necessary material. (F8)

Another two families mentioned that their children only participate in recreational and leisure activities with the school or institution where they spend most of their time, given that they do not have the economic conditions to provide these activities to their children out of school, as can be observed in the following quotes.

> There are many economic difficulties; it is difficult to do activities with our daughter. The few activities that she does are in the school: swimming and gymnastics. And we do not carry out tourism trips as a family. It is unthinkable! (F10)

> We do not participate in tourism activities, only the activities that our son does in the school. It is all so expensive. We prefer to spend the money on the treatments that he needs. (F4)

The results obtained clearly show that access to tourism activities in this group is relatively low. Almost a third of the families surveyed do not carry out tourism activities as a family. This situation is strongly influenced by the great economic constraints of these families, as suggested by Giulio *et al.* (2014). For LIFWCD, in the majority of the cases, it is impossible to engage in recreational and leisure activities, including tourism activities. The results obtained in this study highlight this situation, given that the economic difficulties are the main factor inhibiting these families from engaging in recreational and leisure activities together with their children, corroborating the results presented in the study of Giulio *et al.* (2014).

The lack of accessible tourism destinations is the second travel constraint (structural constraint) most mentioned by the families interviewed in this study. Neither facilities nor human resources are prepared to provide activities for children with disabilities, corroborating the results of other studies (e.g. Kim and Lehto, 2013).

The results obtained in this research allow us to conclude that, although the practice of tourism activities is not a reality for all families surveyed, mainly owing to economic and structural constraints, families perceived various benefits of the tourism activities carried out as a family both for the children and the whole family. One of the main benefits mentioned is related to the improvement of family ties, as can be observed in the following quotes.

> We were able to participate and live more with our son, it is very good! And we feel stronger family ties. (F1)
>
> These are very rewarding moments in the family! We have very good experiences! (F5)
>
> A lot of family fun! It is good to keep up with the growth of [son], especially to overcome his difficulties. (F6)
>
> It is very good for us, because we have a good family time. (F11)

Personal development, particularly of the children with disabilities, was also one of the benefits frequently mentioned by the families interviewed. This is clear in the following quotes:

> It is good to show the world that all of us are capable, but with different timings. (F6)
>
> The benefits are huge! Relaxation, joint growth, the possibility of exploring the environment, sharing knowledge and stimulating the [child] at a visual level. (F7)

The satisfaction of the families interviewed in carrying out recreation and leisure activities together with their children is notable. These families consider that these activities have a crucial importance in the development of their children and in the strengthening of family ties. These results are in accordance with the results of other studies in this field (e.g. Mactavish and Schleien, 1998; Kim and Lehto, 2013). However, although the majority of the families interviewed identified several benefits of tourism trips carried out as a family, one of the families mentioned that they do not obtain benefits from the leisure and recreation activities carried out with their children. They consider that the family is restricted from visiting certain places because of the condition of the child, given that several leisure and recreation products are not accessible yet and they need to overcome several barriers. Moreover, they also consider that the child does not benefit from these activities. Therefore, to facilitate the participation of these families in tourism activities it is of utmost importance to overcome the various travel constraints that these families face when planning and carrying out a tourism trip and also to improve the quality of tourism products offered, as suggested by Yau *et al.* (2004) and Kim and Lehto (2013).

According to the families' view, the accessibility of tourism destinations is one of the most important factors that influence the selection of a tourism destination, especially in cases where it is necessary to move the child with a wheelchair or with support. Moreover, these families also prefer quiet, sparsely populated and flat places, where movement is facilitated. Finally, those interviewed mentioned that they take the normal care that all families do when travelling with children, such as noting the proximity of health centres or hospitals and toilets. The importance of the factors mentioned is proved in the following quote.

> We assess if there are a lot of people, if you have bathrooms nearby, because [the child], despite not wearing a diaper, does not ask to pee and so from time to time we have to check ... if there are places where he can rest because he tires faster than we do. (F6)

7.4.4 STP for LIFWCD: Relevance and characteristics

According to the social organizations' view and families' view, LIFWCD desire to participate in tourism activities and perceive various benefits of participation in these activities. However, several LIFWCD do not have the possibility to participate in tourism activities owing to the various intrapersonal and structural constraints that they face. To overcome these constraints, it is of the utmost importance to promote STI to facilitate the participation of these families in tourism activities, promoting 'Tourism for All'. To analyse the relevance of STP in facilitating the access to tourism activities for LIFWCD, the concept of social tourism was explained to each family interviewed. Of the 11 families interviewed, only one vaguely knew the concept. After the explanation of the concept, only two families mentioned knowing of STP for children with disabilities. These results show clearly that these families do not know the concept of social tourism and that they do not recognize the existence of STP that can facilitate their access to tourism products. This result may be related to the limited offer of STP in Portugal in promoting the participation of these families in tourism activities. When the families were questioned regarding whether they liked participating in STP, all families responded affirmatively, showing enthusiasm for this possibility. However, most of those interviewed mentioned that STP should be adapted to the disabilities of the children and should also involve the participation of children without disabilities. These opinions can be corroborated in the following quotes:

> Although we know a social programme for families with children with disabilities, we never participated, because we notice that our son feels better when in contact with children without disabilities. (F1)
>
> We would participate in a social programme if it includes also children without disabilities. Because my husband considers it as exclusion if these programmes include only children with disabilities. (F6)

According to the mother interviewed of family F6, the opinion of her husband is a common opinion of a great number of the families with children with trisomy 21, as can be seen in the following quote:

> My husband's way of thinking is common in many of the families with children with trisomy 21 that we know, especially in the case of the fathers. Mothers accept it better than fathers do. Many fathers do not want to go to meetings, because they think that we ourselves are practising exclusion. (F6).

The families interviewed were also questioned regarding the importance of STP in facilitating the participation of LIFWCD in tourism activities. All the families consider these programmes important, as can be perceived in the following quotes.

> I think that these programmes are very important. They are a benefit to many families. (F1)
>
> Many families do not have the economic capacity to practise tourism or other recreational and leisure activities. Therefore, I consider these programmes very important. (F7)
>
> I think that these programmes are very important because the majority of families with children with disabilities have economic difficulties. And the few economic resources that they have they use for health treatments and other therapies that their children need. These therapies are very expensive. (F6)
>
> It is very important that there are programmes of this type, because families also need to socialize and exchange opinions. (F11)

The results of this study show clearly that LIFWCD face various economic constraints that make access to tourism activities difficult for them. This highlights the importance of STP in overcoming some of these constraints and providing the opportunity for these families to have access to activities that allow them to socialize, have moments of relaxation and escape from all the problems that they have to face in their daily life.

7.5 Conclusions and Implications

The results of this study are in line with other studies (e.g. Mactavish and Schleien, 1998; Forsyth and Jarvis, 2002; Kim and Lehto, 2013; Nyman *et al.*, 2018) that have examined the benefits that families with children with disabilities obtain from participation in tourism activities. The contribution to the improved quality of life of these families, the development of personal and social competences, the improvement of physical and mental health of their children, and the strengthening of family ties are the benefits most identified by the social organizations and families interviewed in this research.

The practice of tourism activities as a family contributes to developing healthier and more stable family relationships. Therefore, it should be of interest to the parents to organize family activities (Kim and Lehto, 2013). The results of the interviews carried out in this research show that one of the most important benefits that the families perceive when tourism activities are done as a family is related to the possibility of all members of the family participating together in the same activities, improving their relationships and reinforcing family ties. Some families also mention that these activities contribute to their children overcoming their difficulties, spending moments of fun and relaxation together. However, there are a great number of LIFWCD that do not participate in tourism activities owing to the various travel constraints that they face, mainly intrapersonal and structural constraints.

Outdoor tourism activities (e.g. walks, going to the beach) and activities that require little physical effort (e.g. going to the cinema and theatre) are the activities most often carried out by the families analysed in this research, which is in line with the results of other studies (e.g. Kim and Lehto, 2013). These activities are chosen taking into account the skills and abilities of their children. The results of this study also show that in most cases, these families seek accessible and calm tourist destinations for the practice of family tourism activities.

This research contributes to increasing knowledge about tourism practices of LIFWCD, the travel constraints that they face and the benefits that these families obtain from tourism activities. These families face many daily challenges for which they need to be prepared, not only physically but also psychologically. The practice of tourism activities as a family will allow these families to acquire energy to face these challenges. However, a great number of these families cannot overcome the travel constraints that they face and consequently do not do tourism activities. As Hazel (2005, p. 225) highlights, 'the provision of holidays away from home for children and families has been recognised as a social policy concern across mainland Europe since the mid-20th century'. However, much more has to be done in terms of social policies in this field. The definition of social policies that promote the participation of LIFWCD in tourism activities is of the utmost importance in promoting a more just and inclusive society.

To increase the participation of LIFWCD in tourism activities, it is important to provide support to these families in order to overcome the great travel constraints that they face, specifically low economic resources and lack of accessible tourism products. Although there are already some STP in Portugal in which these families can participate, the number of these programmes is limited and the majority of these families are unaware of the existence of these programmes (Lima, 2015; Pedrosa, 2016). The results of this research corroborate this situation. The majority of the families interviewed do not know of STP where they can participate. However, after the concept of STP was explained to the families interviewed, all considered that these programmes are extremely relevant.

The families interviewed mentioned that they would like to participate in STP. However, they consider that these programmes should not be exclusive to families with children with disabilities. These programmes should include children with disabilities and children without disabilities to promote inclusion and simultaneously socialization between children with different needs, desires and characteristics. Everybody (children with disabilities, children without disabilities and their families) can benefit from this interaction and sharing of experiences. Therefore, the results of this study highlight the need for developing STP that promote the participation in tourism activities of these families together with families whose children are not disabled. It is crucial that disabled children and their families do not feel excluded.

Despite the contributions of this chapter to increasing knowledge of tourism practices of a market segment that has been neglected in the literature, this research does have some limitations. First, this is an exploratory study. The sample of families as well as the sample of social organizations that participated in this research are too small to draw general conclusions about the tourism practices of these families, the benefits obtained from tourism activities and also the types of STP that can be developed to increase the participation of these families in tourism activities. However, the samples are large enough to gain an exploratory understanding regarding the topics under analysis. To increase the research on this topic and to define social policies that promote the adoption of the philosophy of 'Tourism for All', it is essential to undertake further research; for example, in terms of the impacts of tourism activities on the well-being of these families, how to design STP for this market that also include the participation of families with children without disabilities, and how the tourism destinations and the human resources of the tourism industry should be prepared to provide services to this market.

Acknowledgements

This work was financially supported by the research unit on Governance, Competitiveness and Public Policy (UID/CPO/04058/2019), funded by national funds through FCT – Fundação para a Ciência e a Tecnologia.

References

Blichfeldt, B.S. and Nicolaisen, J. (2011) Disabled travel: Not easy, but doable. *Current Issues in Tourism* 14(1), 79–102.

Carneiro, M.J., Eusébio, C., Kastenholz, E. and Alvelos, H. (2013) Motivations to participate in social tourism programmes: A segmentation analysis of the senior market. *Anatolia: An International Journal of Tourism and Hospitality Research* 24(3), 352–366.

Darcy, S., Cameron, B. and Pegg, S. (2010) Accessible tourism and sustainability: A discussion and case study. *Journal of Sustainable Tourism* 18(4), 515–537.

Deville, E. and Kastenholz, E. (2018) Accessible tourism experiences: The voice of people with visual disabilities. *Journal of Policy Research in Tourism, Leisure and Events* 10(3), 265–285.

Diekmann, A. and McCabe, S. (2011) Systems of social tourism in the European Union: A critical review. *Current Issues in Tourism* 14(5), 417–430.

Dobson, B. and Middleton, S. (1998) *Paying to Care: The Cost of Childhood Disability*. York Publishing Services, New York.

Dowling, M. and Dolan, L. (2010) Families with children with disabilities – inequalities and the social model. *Disability & Society* 16(1), 21–35.

Eusébio, C. and Carneiro, M.J. (2011) Determinants of tourism impact on tourists' quality of life. *International Journal of Leisure and Tourism Marketing* 2(4), 313–336.

Eusébio, C. and Carneiro, M.J. (2014) The impact of tourism on quality of life: A segmentation analysis of the youth market. *Tourism Analysis* 19(6), 741–757.

Eusébio, C., Carneiro, M.J., Kastenholz, E. and Alvelos, H. (2013) The socioeconomic impacts of INATEL social tourism programs for the senior market. In: Diekmann, A. and Jolin, L. (eds) *Social Tourism: International Perspectives the Contribution of Research*. Presses de l' Université du Québec, Québec, Canada, pp. 197–213.

Ferrer, J.G., Sanz, M.F., Ferrandis, E.D., McCabe, S. and García, J.S. (2016) Social tourism and healthy ageing. *International Journal of Tourism Research* 18, 297–307.

Figueiredo, E., Eusébio, C. and Kastenholz, E. (2012) How diverse are tourists with disabilities? A pilot study on accessible leisure tourism experiences in Portugal. *International Journal of Tourism Research* 14(6), 531–550.

Fontes, F., Martins, B.S. and Hespanha, P. (2014) The emancipation of disability studies in Portugal. *Disability & Society* 29(6), 849–862.

Forsyth, R. and Jarvis, S. (2002) Participation in childhood. *Child: Care, Health and Development* 28(4), 277–279.

Giulio, P., Philipov, D. and Jaschinski, I. (2014) Families with disabled children in different European countries. Families and societies working paper series, 23. Available at: www.familiesandsocieties.eu/wp-content/uploads/2014/12/WP23GiulioEtAl.pdf (accessed 15 October 2020).

Glidden, L. (1993) What we do not know about families with children who have developmental disabilities: Questionnaire on resources and stress as a case study. *American Journal on Mental Retardation* 93(5), 481–495.

Hazel, N. (2005) Holidays for children and families in need: An exploration of the research and policy context for social tourism in the UK. *Children & Society* 19, 225–236.

Higgins-Desbiolles, F. (2006) More than an 'industry': The forgotten power of tourism as a social force. *Tourism Management* 27(6), 1192–1208.

Kastenholz, E., Eusébio, C., Figueiredo, E. and Lima, J. (2012) Accessibility as competitive advantage of a tourism destination: the case of Lousã. In: Hyde, K.F., Ryan, C. and Woodside, A.G. (eds) *Field Guide to Case Study Research in Tourism, Hospitality and Leisure: Advances in Culture,*

Tourism and Hospitality Research. Volume 6. Emerald Group Publishing Limited, Bingley, UK, pp. 369–385.
Kastenholz, E., Eusébio, C. and Figueiredo, E. (2015) Contributions of tourism to social inclusion of persons with disability. *Disability and Society* 30, 1259–1281.
Kim, S. and Lehto, X.Y. (2013) Travel by families with children possessing disabilities: Motives and activities. *Tourism Management* 37, 13–24.
Larson, R. and Verma, S. (1999) How children and adolescents spend time across cultural settings of the world: Work, play and developmental opportunities. *Psychological Bulletin* 125(6), 701–736.
Lima, J. (2015) Turismo em família: a importância do turismo para famílias economicamente carenciadas. [Family tourism: the importance of tourism for low-income families.]. PhD dissertation. University of Aveiro, Aveiro.
Lima, J., Eusébio, C. and Varum, C.A. (2011) O combate exclusão social através de programas de turismo social para famílias economicamente carenciadas. [Tackling social exclusion through social tourism programs for economically disadvantaged families]. *International Conference on Tourism & Management Studies* 1, 256–268. Available at: http://hdl.handle.net/10773/6532
Lima, J., Eusébio, C. and Varum, C.A. (2012) Effects of holidays on tourists: A review and directions for future research. *RT&D/Journal of Tourism and Development 17/* 18(1), 265–277.
Mactavish, J. and Schleien, S. (1998) Playing together growing together: Parents perspectives on the benefits of family recreation in families that include children with a development disability. *Therapeutic Recreation Journal* 32(3), 207–230.
Mactavish, J.B. and Schleien, S.J. (2004) Re-injecting spontaneity and balance in family life: Parents' perspectives on recreation in families that include children with developmental disability. *Journal of Intellectual Disability Research* 48(2), 123–141.
McCabe, S. (2009) Who needs a holiday? Evaluating social tourism. *Annals of Tourism Research* 36(4), 667–688. DOI: 10.1016/j.annals.2009.06.005.
McCabe, S., Minnaert, L. and Diekmann, A. (2011) *Social Tourism in Europe – Theory and Practice*. Channel View Publications, Bristol, UK.
Minnaert, L. (2014) Social tourism participation: The role of tourism inexperience and uncertainty. *Tourism Management* 40, 282–289.
Minnaert, L. and Shapmans, M. (2009) Tourism as a form of social intervention: The holiday participation centre in Flanders. *Journal of Social Intervention: Theory and Practice* 18(3), 42–61.
Minnaert, L., Maitland, R. and Miller, G. (2009) Tourism and social policy – the value of social tourism. *Annals of Tourism Research* 36(2), 316–334.
Minnaert, L., Maitland, R. and Miller, G. (2011) What is social tourism? *Current Issues in Tourism* 14(5), 403–415.
Mitchell, R.D. (1998) Learning through play and pleasure travel: Using play literature to enhance research into touristic learning. *Current Issues in Tourism* 1, 176–188.
Moura, A.E.A., Kastenholz, E. and Pereira, A.M.S. (2018) Accessible tourism and its benefits for coping with stress. *Journal of Policy Research in Tourism, Leisure and Events* 10(3), 265–285.
Murray, M. and Sproats, J. (1990) The disabled traveller: Tourism and disability in Australia. *The Journal of Tourism Studies* 1(1), 9–14.
Nyman, E., Westin, K. and Carson, D. (2018) Tourism destination choice sets for families with wheelchair-bound children. *Tourism Recreation Research* 43(1), 26–38.
Oldman, C. and Beresford, B. (1998) *Homes Units for Children: Housing, Disabled Children and Their Families*. Policy Press, Bristol, UK.
Pedrosa, M. (2016) Turismo social para famílias com crianças ou jovens com incapacidade: Proposta de programas. [Social tourism for families with children or young people with disabilities: Program proposal.]. Master's thesis. University of Aveiro, Aveiro.
Rickford, F. (2000) The forgotten families. *Community Care* 6, 18–20.
Smith, R. (1987) Leisure of disabled tourists: Barriers to participation. *Annals of Tourism Research* 14, 376–389.
Smith, V. and Hughes, H. (1999) Disadvantaged families and the meaning of the holiday. *International Journal of Tourism Research* 1, 123–133.
Townsend, J., Zabriskie, A. and Ramon, B. (2010) Family leisure among families with a child in mental health treatment: therapeutic recreation implications. *Therapeutic Recreation Journal* XLIV(1), 11–34.

United Nations (1948) Declaração universal DOS Direitos Humanos. Available at: http://unesdoc.unesco.org/images/0013/001394/139423por.pdf (accessed 15 October 2020).

Valentine, F. (2001) Enabling citizenship: full inclusion of children with disabilities and parents. Canadian policy research networks (CPRN) discussion paper No. F/13. Available at: http://cprn.org/documents/ACFZwv9Kd.PDF (accessed 15 October 2020).

Yau, M.K., McKercher, B. and Packer, T.L. (2004) Traveling with a disability. *Annals of Tourism Research* 31(4), 946–960.

8 Disabled Children as Legitimate Research Participants: A Topic Omitted from Tourism Research?

Fathimath Shiraani[1]*, Neil Carr[1] and Gill Rutherford[2]
[1]Department of Tourism, University of Otago, Dunedin, New Zealand; [2]College of Education, University of Otago, Dunedin, New Zealand

8.1 Introduction

Tourism research is following other disciplinary fields in removing the invisible 'walls' blocking children's/young people's (here onwards children[1]) agency, access to participation and voice. Research with children is emerging in the field, but one needs to question tourism researchers' orientation and approach towards children. To what extent are tourism scholars contemplating the diversity of children? Are they privileging some groups of children over others? Have they left behind children cohorts, like disabled children? While recognizing the significance and breadth of such questions, this chapter focuses on disabled children from a tourism perspective.

Some may have reservations about identifying children as 'disabled children' or associating children with the word 'disability'. However, without this acknowledgement, disabled children's presence may be neglected. In the same vein, what disabled children share may become superficial, less visible and lost when they are classified under labels such as children or disabled people per se. Breaking the silence around disabled children's inclusion and exclusion, particularly in relation to tourism, is essential, considering the growth in the global disabled children's population (Reiner *et al.*, 2019) and the potential benefits for disabled children and their families of engaging in tourism experiences (Kim and Lehto, 2013; Cloquet *et al.*, 2018).

Consequently, the aim of this chapter is to provide a discussion of the need to respect and listen to the views of disabled children in general, and in relation to tourism specifically. This discussion is positioned within the broader debate on children's agency and rights to voice. As part of this, the chapter explores why obtaining children's, including disabled children's, perspectives is invaluable to understand the matters that are important to them. Recognizing the voice of disabled children and actively listening to them is important to aid the promotion of children rights (Ajodhia-Andrews, 2016). Despite the importance of disabled children's voices, the chapter argues that they are underrepresented in tourism scholarship.

The chapter starts by defining the concept of voice and summarizes the broader debates and arguments regarding children's voice. The chapter then continues to cover how some children's voices, referred to as 'conventional voices' by Richards *et al.* (2015, p. 121) are privileged over others (such as disabled children). The

*Corresponding author: fathimath.shiraani@postgrad.otago.ac.nz

 Social Tourism: Global Challenges and Approaches (eds J. Lima and C. Eusébio)
DOI: 10.1079/9781789241211.0009

subsequent section reflects on two strands of tourism research related to disability and children and highlights disabled children's invisibility and omission of their perspectives in tourism research. The chapter then provides a brief discussion of the challenges of researching with disabled children and outlines some ethical and practical considerations. Overall, this chapter serves as a springboard for further inquiry concerning disabled children's inclusion in tourism research.

8.2 Why is Listening to Children's Voices So Important?

Voice has become a burgeoning mantra in research and practice. A literal or dictionary definition of voice suggests it is spoken words or speech. This restrictive understanding may not be acceptable to many as voice as a pronounced 'metaphor' (e.g. Schnoor, 2012) is socially constructed in numerous ways (Ajodhia-Andrews, 2016). Within this context, there is no standard agreement on the meaning of voice. In this chapter, the notion of children's 'voice' refers to the unadulterated perspectives, views, thoughts (Messiou, 2006), 'intentions, hopes, grievances, and expectations' communicated by children themselves (Pufall and Unsworth, 2004, p. 8).

Children's voices are valued as 'credible and worthy of being listened to' (Richards *et al.*, 2015, p. 121) and welcoming these voices is seen as a way to liberate and represent children on equal terms with other groups in society (Qvortrup, 2015). While other similar benefits are reported (e.g. personal growth of children and better decisions and outcomes with regard to children's legislations, policies and programmes), there remains considerable resistance to listening to the voice of children (Lansdown, 2011). For example, it has been suggested that children lack competence to voice their opinions or that such communication could curtail children's childhood (Lansdown, 2011). Regardless of the latter, children's perspectives and experiences are increasingly used to inform policy, practice and research.

Many factors have added to the increasing prominence of valuing what children have to communicate. One major factor that drove worldwide interest was the enactment of the United Nations Convention on the Rights of the Child (UNCRC) in 1989 (United Nations, 1989). Article 12 of the UNCRC requires Member States to afford *all* children the right of expression, to be heard and to be considered seriously (United Nations, 1989). The ratification of the UNCRC was a watershed event for children's voice, but it is important to note that the UNCRC occurred alongside other developments. One such development was an ideological shift in childhood research that led to the recognition of children's agency and reconstruction of children as social actors (Mason and Hood, 2011). Before this conceptual reconstruction, children were characterized as passive, incompetent and incomplete objects in need of adult protection (Prout and James, 2015). The theoretical transformation from 'child as not yet a person, [to] the child as person and the child as agent' (O'Neill, 2014, p. 219) lifted the status of children from passive to active actors.

Coyne and Carter (2018) identified three additional developments that triggered recognition to children's voice. First, there were legislative acts in many countries supporting the inclusion of children's voice in matters relevant to them. Second, increasingly child-friendly legislative demands require including children's (for example, as users of education and health care) viewpoints on various policies, reports and guidelines. Third, political and funding bodies have advocated for engagement with children in research concerning them.

Research conveying children's perspectives on various matters, such as education (e.g. Sargeant and Gillett-Swan, 2015), health (e.g. Blades *et al.*, 2013), and child protection and welfare (e.g. Cossar *et al.*, 2016) is increasing with due weight given to uphold and respect children's perspectives. However, an important question to ask is whether these studies represent all children? Similarly, do all children get an equal chance to share their experience?

Regrettably, selective children tend to get the opportunity to be heard while others remain unheard (Tisdall, 2017). Richards *et al.* (2015) argue that voices that are easier to be heard and interpreted, referred to as conventional voices, are privileged, sought after, and therefore heard. Conversely, they contend that harder to hear voices (unconventional voices) are either neglected or silenced. Richards *et al.* (2015) suggest that the unconventional voices should be prioritized and the agentic status of these children should be upheld to gain a holistic representation of children's voices.

8.3 Disabled Children's Voices

The right to voice enshrined in the UNCRC is meant to apply to all children. It indicates disabled children have equal rights to be included and heard. Disabled children's right to voice is further strengthened in Article 7 of the United Nations Convention on the Rights of Persons with Disabilities – UNCRPD (United Nations, 2006). Despite these conventions it is commonly claimed that 'disabled children are excluded as participants from almost all research' (UNICEF, 2013; Roberts, 2017, p. 153). Consequently, comparative to other children, very little is known or reported about disabled children's views. So why is there an apparent reluctance to include disabled children in research?

Limited acceptance of children's diversity is one significant barrier hindering inclusion of disabled children in research. This is related to the tradition of viewing children as a homogeneous group (Stafford, 2017). Homogenization has led to 'othering' of children who are perceived to be outside the boundaries of the 'normal' child. Disabled children are commonly placed outside the so-called normal child frame and regarded as the 'other' to be excluded from childhood research (Runswick-Cole *et al.*, 2018a). Disabled children's inclusion in research was also restricted due to the imposition of such labels as 'sick child', 'abnormal child' and other deficit descriptors (Cooper, 2013). Such conformist views and orthodox thinking negatively identified disabled children as objects only of medical research. The areas of research assigned as appropriate for disabled children have thus been restricted, with their lives heavily studied from medical ideologies of disability (Runswick-Cole *et al.*, 2018a). As objects of research, disabled children were silenced, with adult proxies acting on their behalf (Watson, 2012).

Evolving work is contesting the homogeneous identity of children and use of proxy informants. Emergent research (e.g. Curran and Runswick-Cole, 2014; Runswick-Cole *et al.*, 2018b) is demanding the exploration of different childhoods and promoting the move beyond medical/impairment-oriented research that neglects disabled children's everyday experiences. Stressing this point, Roberts (2017, p. 153) writes that 'disabled children and young people know about issues other than their impairment'.

While discussions about children's diversity continue, it is vital to realize that disabled children as a group are not identical (Conder *et al.*, 2016). As noted by Matthews (2009), appreciating disabled children's diversity requires moving beyond the image of those using wheelchairs. Disabled children constitute children of different demographic characteristics (e.g. age, gender and race) who have various impairments (e.g. physical, intellectual or sensory impairments) (see UNCRPD, Article 1; United Nations, 2006) that may be visible and/or invisible.

Another significant barrier faced by disabled children and children, in general, is societal assumptions about their capacity and competence (Stafford, 2017). These assumptions see children, and particularly disabled children often judged by others to be incapable, immature, vulnerable and/or incompetent to participate in research (Richards *et al.*, 2015). According to Stafford (2017), such assumptions are based on constructs of ableism and adultism. Ableism reflects notions of normalcy and able-bodiedness. Adultism refers to the negative power exerted by adults over children owing to perceptions about their vulnerability (Stafford, 2017).

However, emerging research (e.g. Hodge and Runswick-Cole, 2013; Pickering, 2018) suggests that disabled children can make meaningful contributions as research participants if the right opportunities are provided. Facilitating research participation as active participants is viewed as being beneficial for disabled children. Some potential benefits of involving disabled children in research are: increase in self-esteem and self-confidence, opportunity to socialize and gain new skills and experiences, sense of greater responsibility and independence, and improved knowledge and access to decision-making (Bailey *et al.*, 2014).

Despite the emergence of the recognition of the active agency of disabled children, some fields (like tourism) lag behind the practice of involving this group of children in research.

8.4 The In/visibility of Disabled Children in Tourism Research

In recent times, tourism scholars have shown a growing interest in various marginalized groups, including children and disabled people.

A group of researchers in the early 1980s and 1990s (e.g. Graburn, 1983; Cullingford, 1995; Thornton *et al.*, 1997) brought attention to children and raised concerns about limited recognition of children in tourism research. Work from the early writers contributed to child-friendly research, but many of the publications were either confined to limited topics (such as how children influenced parental decision-making or tourist experiences) or presented from adult perspectives (Small, 2008). As such, children were not given an opportunity to share their worlds and, instead, selected adults spoke on behalf of the children.

Particularly the idea of working with children to capture their viewpoints grew from the work of researchers such as Cullingford (1995), Nickerson and Jurowski (2001) and Gram (2005). While these studies did make a point about children's inclusion and voice, many of them were economically driven to present children as a vital market segment (Small, 2008). Irrespective of the motives, these studies contested earlier assumptions about children as the silent property of parents/whānau[2]/caregivers – here onwards referred to as parents or guardians (Singh and Nayak, 2016). This has led to the gradual growth in research actively seeking children's views (Canosa and Graham, 2016). Alongside this has been the development of family tourism research that has adopted a whole-family approach. Whole-family research recognizes children as active and influential consumers/participants within the family unit (Schänzel, 2010; Schänzel and Yeoman, 2015). The shift towards tourism researchers recognizing the active agency of children has reflected and been driven by this earlier recognition in other fields (e.g. childhood studies) (Carr, 2011; Canosa and Graham, 2016).

Like children, disabled people began to be studied by tourism researchers around the 1980s and early 1990s (Yau *et al.*, 2004). The early scholars (like Smith, 1987; Murray and Sproats, 1990) identified disabled people as a potential niche market and argued that disabled people's participation in tourism was constrained by multiple barriers (e.g. economic, physical and attitudinal barriers).

Disability discourse has filtered into various strands of tourism research such as hopeful tourism (e.g. Richards *et al.*, 2010), social tourism (e.g. Zenko and Sardi, 2014) and accessible tourism (Darcy, 2006). While the initial research was focused on understanding and tackling barriers and access issues faced by disabled people, disability-related tourism research is now evolving to capture more complex understandings of disability (Buhalis and Darcy, 2011; Michopoulou *et al.*, 2015). Within this context, tourism scholars are realizing that a universal one-size-fits-all solution to tourism access for disabled people does not work (McKercher and Darcy, 2018) as disabled people do not constitute a group of homogeneous people (Figueiredo *et al.*, 2012). Based on this understanding, tourism scholars are increasingly engaging with different cohorts of disabled people – for instance, those with visual impairments (e.g. Loi and Kong, 2017), hearing impairments (e.g. Ho and Peng, 2017), mental health issues (e.g. Page *et al.*, 2015) and intellectual disabilities (e.g. Sedgley *et al.*, 2017). The research is also shifting from a market-dominated focus to one that is person-centred.

As highlighted in Lehto *et al.* (2018, p. 186) disabled people's perspectives about tourism have been collected on diverse topics (e.g. tourism/travel motivation, benefits of tourism) and on various industry products and services such as 'in-flight, museum, hotel, ferry, and restaurant'. However, this body of literature is facing criticism for undervaluing the role of family (Cloquet *et al.*, 2018) and caregivers (Lehto *et al.*, 2018) in disabled people's tourism experiences although they are typically accompanied by family members/caregivers (Figueiredo *et al.*, 2012; Kim and Lehto, 2013). The omission of family is surprising considering the advancement in family tourism and the utilization of the whole-family approach within tourism research.

The lack of interchange between disability and family tourism research is highlighted in the work of Cloquet *et al.* (2018), who have called for a whole-family perspective to disability in tourism research. This chapter is supportive of this call and stresses that such an approach is vital to balance the current overrepresentation of disabled single adults and underrepresentation of disabled children in current disability-related tourism research. It also offers the opportunity to position disabled children in family tourism research.

Tourism research with children is generally silent on the topic of disabled children. Equally, tourism research with disabled people has tended to ignore disabled children. Only a few studies (e.g. Kim and Lehto, 2013; Sedgley *et al.*, 2017; Nyman *et al.*, 2018) buck this trend and even these have failed to recognize their agency and rights to voice. Instead, these children are represented in a superficial manner through proxy informants. As mentioned earlier in this chapter, the use of proxies to study children/disabled children has been contested in contemporary childhood research. Substituting disabled children with proxies does not sit well with disability advocacy efforts.

Even in social tourism research, as an area of research that is accommodative of various marginalized groups' rights to tourism (Minnaert *et al.*, 2006), disabled children's voices if heard at all tend to be mixed with those of other marginalized groups (e.g. children from low-income backgrounds/families) (see McCabe, 2009).

By paying more attention to the voices of disabled children we can learn about their tourism experiences. This will aid the development of fully informed inclusive tourism policies and practices. Developing such policies/practices is important given the recognition that there are more than 93 million disabled children in the world (UNICEF, 2018), representing one in every ten children (UNICEF, 2017). Given the scale of this group, understanding tourism experiences of disabled children is important for the tourism industry, in terms of product development, customer satisfaction and brand development. Listening to the voices of disabled children may also be fruitful for disability advocacy and disability service organizations assisting with disabled children's tourism experiences.

While research dedicated to the voices of disabled children is needed, it is equally clear that this needs to be set not just in a whole-family setting but in a broader social context. This is needed in order to recognize the wider situation in which disabled people live and to avoid marginalization of these groups of people.

8.5 Researching with Disabled Children

Once the need to listen to the voices of disabled children is recognized, consideration must be given as to how this should be done. In conducting research with vulnerable groups, ethical considerations take centre stage. It is necessary to have an ongoing responsive approach to ethics (Wickenden and Kembhavi-Tam, 2014) and to consider both procedural ethics and ethics in practice. Procedural ethics refers to obtaining ethical approval from ethics committees or institutional ethics review boards, whereas ethics in practice covers various considerations that should be made as the research proceeds and develops in the field (Phelan and Kinsella, 2013). Key concerns of procedural ethics and some issues pertinent to ethics in practice are addressed below.

8.5.1 Procedural ethics – ethical approval

Seeking approval from ethical regulatory boards is common practice in research involving human participants. Ethics committees are particularly concerned with ensuring the welfare of children (Richards and Clark, 2018). This protectionist rationale is doubly strong when research is conducted with disabled children (Loveridge and Cornforth, 2014).

Protectionist thinking puts disabled children in a paternalistic, vulnerable frame. A vulnerable image is problematic as it may influence ethical review committees' decision-making and the type of protective measures they demand (Skyrme and Woods, 2018). Ethics committees may use more stringent surveillance and safeguards (Graham and Fitzgerald, 2010) as the term vulnerability is too often 'seen as interchangeable with the notion of lacking competence' (Carter, 2009, p. 861). An overcautious approach may compete with the ideologies about children's agency and participation. As argued by Powell and Smith (2009), protectionist ideologies can repudiate children's right to participate. Tisdall (2017) similarly adds that children's participation rights can be 'blocked' if they are perceived to be too vulnerable. Thus, navigating contending ideologies about protection and participation is a complex reality for researchers working with children. Researchers have to be vigilant that protectionist ideologies or stereotypes are not used by ethics committees as a weapon to exclude or avoid researching with disabled children. Equally, the vested interest of the researcher to involve and represent disabled children must not be tokenistic or exploitative. Furthermore, a desire to enable the participation of children in research should not be used as an excuse to loosen measures designed to protect them. Being open to all these possibilities (protection, exploitation and participation) is essential for researchers working with children/disabled children (Tisdall, 2017). The possibilities should be recognized and reflected on not only when pursuing ethical approval but also when in the field.

8.5.2 Ethical concerns in practice

There are various ethical concerns that researchers should be aware of when in the field. Some include negotiating with multiple gatekeepers, seeking informed consent and addressing power imbalances.

Working with gatekeepers

Ethical approval does not guarantee direct access to disabled children. The opportunity to reach disabled children depends on the willingness and sponsorship of a variety of gatekeepers. Gatekeepers refer to those who 'control or limit researchers' access to the participants' (Coyne, 2010a, p. 452). Gatekeepers include parents, caregivers, hospitals and health professionals, schools and other community services/organizations (Coyne, 2010a; Moore *et al.*, 2011). Gatekeepers see themselves as being responsible for protecting research participants from exploitation, invasion, distress and/or harm (Coyne, 2010a).

Like members of an ethics committee, gatekeepers may inhibit or facilitate access to disabled children (Richards and Clark, 2018). The reaction of gatekeepers depends on their belief systems and understanding of the research. Any negative judgements about the research (e.g. research is an additional source of stress; it is a waste of time), the researcher (e.g. the researcher cannot be trusted, he/she is an intruder/stranger) and/or about the child's suitability or competency (e.g. they would not know) can block access to children/disabled children (Coyne, 2010a; Moore *et al.*, 2011). Wickenden and Kembhavi-Tam (2014) specifically highlighted why disabled children's parents may be reluctant to approve their children's involvement in research. Reasons include fear of abuse, concern about confidentiality and the privacy of their children, and lack of assurance from researchers that activities/methods will be tailored and adapted according to their child's needs. Gatekeepers' lack of understanding of their role in the research process may also create a barrier. All of this places the onus on the researcher to provide a clear and concise explanation of the study. Inadequate or misleading explanations of a study provided by gatekeepers to disabled children could result in a child declining to be involved in a study (Coyne, 2010a).

Conversely, gatekeepers can be cooperative and helpful in connecting researchers with children. Gatekeepers can act as a research facilitator in numerous ways, including scheduling appropriate contact times/places, introducing the study, breaking the ice and putting the child at ease (Coyne, 2010a; Abbott, 2013). They may also act as a trusted and supportive third party (e.g. as an interpreter) (Davis *et al.*, 2017). Particularly for studies with disabled children, gatekeepers' collaboration is crucial to gain knowledge on how to adapt and modify the consent process and data collection methods/processes according to the requirements or routines of the child (Shaw *et al.*, 2011). Similarly, information from gatekeepers can help to ascertain whether study participants are capable of consenting on their own without the provision of others (Jenkin *et al.*, 2015). It is imperative for researchers to be cautious of the power difference between the child and the gatekeepers, as discussed later in the chapter. In this regard, researchers need to make sure that child participants are not pressured into participation by gatekeepers (Holland *et al.*, 2010).

Given the centrality of gatekeepers to research with disabled children, researchers must deliver sufficient, useful and easy-to-understand information to them so that the purpose and benefits of the research is transparent (Richards and Clark, 2018). Engaging and seeking gatekeepers' input/viewpoints and reflecting on their concerns is a vital part of establishing trust and support between them and the researcher.

Informed consent

For research with children, obtaining consent from parents/caregivers and children is standard required practice (Jenkin *et al.*, 2015). Consent is an explicit voluntary agreement that research participants can give having been informed of and having understood the nature of the research (Graham *et al.*, 2013). Instead of consent, some researchers may seek assent from children. Assent is viewed as an informal agreement (Graham *et al.*, 2013) with 'spurious quasi-legal status' (Alderson and Morrow, 2011, p. 104). In contrast, consent is referred to as a formal agreement with legal rights and comes with two requirements: to demonstrate competency and understanding of what is involved, and make an informed judgement to participate or dissent – refuse and withdraw (Coyne, 2010b). The debate regarding whether researchers should opt for gaining consent or assent is unresolved.

Obtaining consent from children is best practice. However, an emphasis on assessing children's competency and understanding is problematic, especially for disabled children (Groundwater-Smith *et al.*, 2015; Stafford, 2017). Often disabled children's capacity to understand (particularly of those children with intellectual disabilities) is questioned, and assumptions are made about their inability to comprehend (Horner-Johnson and Bailey, 2013; Stafford, 2017). Researchers should avoid falling into the trap of making quick judgements based solely on 'disability' labels. It is vital for researchers at all times to be responsive to children's diversity and evolving capacities. It is the responsibility of researchers to create accessible and child-friendly consent processes that are considerate of individual children's capabilities and comprehension.

A carefully designed consent process will address aspects such as when and how to obtain consent. An ongoing consent approach rather than a one-off approach is recommended (Alderson and Morrow, 2011). An ongoing process allows children to exercise their right to refuse or continue participation at any point in time during the research. This requires researchers to be observant of verbal, non-verbal and behavioural cues, and to reflect sensitively to recognize freely given consent and dissent. Observation is crucial, as some children may shy away from making an explicit spoken request when it comes to withdrawal from research (Spriggs, 2010).

The quality of the consent process depends on the type of materials and strategies utilized. Employing flexible, child-centred consent materials and strategies is essential, as the standard adult-centric written consent forms may be intimidating for some children. For instance, children with cognitive and intellectual impairments may find difficulty in reading and/or understanding information in written form (Health and Disability Ethics Committees, 2018). In such cases, an oral consent procedure and strategies will be more appropriate (Cameron and Murphy, 2006; Horner-Johnson and Bailey, 2013).

The necessity and pursuit of parental/guardian consent can be controversial, involving discussions about protection, voice and competency. For some academics, parental consent is unacceptable as it promotes deficit thinking (e.g. children are incapable of deciding for themselves) about children's agency and capability (Carr, 2011). Parental consent is similarly criticized as it could act as a barrier inhibiting children's participation if parents refuse to give consent (Coyne, 2010b). Those sceptical about parental consent question whether it is necessary for all children (e.g. older children) and all types of research (e.g. less risky research) (Alderson and Morrow, 2011; Jenkin *et al.*, 2015). Ascertaining children's maturity and/or the risks entailed in a research project is a difficult task (Alderson and Morrow, 2011). Thus, proponents of parental consent believe that involving parents is in the best interest of children (Graham *et al.*, 2013). With arguments for and against parental consent, researchers are left in a paradoxical situation to consider whether permission from parents is necessary or not.

Attention has been drawn to multiple issues that researchers should critically reflect upon when seeking consent from children and/or parents. Additional complexity can surface if the interest of the parents clashes with that of children or vice versa (Graham *et al.*, 2013). If a mismatch happens (i.e. parental consent is obtained in the absence of children's consent, or children's approval is received without parental permission), the researcher will be coiled in a tricky situation – whether to privilege parents' or children's voices. There are no clear-cut solutions to manage such situations. Therefore, actions should be taken after evaluating the situation on a case-by-case basis. In each case, researchers should be vigilant of the power imbalance between the child, parents and the researcher.

Power imbalances

One of the crucial issues to consider when researching with children is the power differential between them and adults. The current trend of researching with children has 'constructed' children as 'powerful' social actors (Richards and Clark, 2018, p. 200) but if and how they can exercise this power within the research setting is related to the power wielded by gatekeepers and researchers.

When researching with children, academics have a responsibility to ensure children are able to utilize their power. This includes in relation to consenting to take part in a study, as noted earlier in the chapter, and when involved in the research. Even if the researcher cannot change the social differences between themselves (as adult and researcher) and the children they are working with, it is possible for them to address their social distance (research relationships) with the children. Some techniques to minimize social distance include: acknowledging and becoming attuned to power inequalities between the researcher and children, delimiting researcher control and being aware of context-specific differences (Raffety, 2015). This could be done, for instance, by adopting a 'least-adult' role as suggested by Mandell (1988), which requires accepting children as they are, suspending any judgements about their capabilities and taking joint actions. Given the power of gatekeepers in general and parents in particular over children it is also vital for researchers to monitor and control this in such a way that children are empowered and protected during the research.

8.6 Conclusions

This chapter has addressed the importance of disabled children's agency, voice and participation in research. In this context, it has emphasized that disabled children are active social agents with a voice. However, there are several obstacles hindering the expression of disabled children's agency and voice. These barriers include a lack of respect for children's diversity, and assumptions about children's capacity and competence. While researchers from other fields (e.g. disabled children's childhood studies) are navigating these obstacles to create space for disabled children, tourism scholarship has rarely considered this group of children and, where it has done, there has been a tendency to view them as objects rather than active agents. Highlighting the dearth of disabled children in tourism research, this chapter has emphasized the need for further research that provides the opportunities and means for disabled children to engage in mainstream tourism research. Disabled children's involvement in research has the potential to aid in designing inclusive tourism policies, products, services and practices that are child-friendly and relevant to this cohort of children. In turn, knowledge from this group has the potential to help inform the work of disability advocates and community organizations. Researching with disabled children opens an opportunity for tourism scholars to engage with and contribute to other fields of inquiry such as disabled children's childhood studies, which calls for the discussion of disabled children within the context of concepts of both childhood and disability. Finally, the chapter addressed some of the practical ethical considerations concerning ethical approval and fieldwork specific to disabled children, many of which ironically have arguably acted as a barrier to those wishing to conduct research on the voices of children in general and disabled children in particular.

Notes

[1] The term 'children' includes all young people of 18 years or below. This definition is utilized by the United Nations Convention on the Rights of the Child (Article 1) (United Nations, 1989). Whether 'disabled children' is the 'right' terminology to use is a matter of debate. This chapter uses disabled children instead of children with disabilities as it reflects the preference of scholars (e.g. Runswick-Cole *et al.*, 2018b) advocating for the agency of disabled children.
[2] Whānau is a Māori-language word for extended family or community of related families.

References

Abbott, D. (2013) Who says what, where, why and how? Doing real-world research with disabled children, young people and family members. In: Curran, T. and Runswick-Cole, K. (eds) *Disabled Children's Childhood Studies: Critical Approaches in a Global Context*. Palgrave Macmillan, Basingstoke, UK, pp. 39–56.

Ajodhia-Andrews, A. (2016) Reflexively conducting research with ethnically diverse children with disabilities. *The Qualitative Report* 21(2), 252–287.

Alderson, P. and Morrow, V. (2011) Consent. In: Alderson, P. and Morrow, V. (eds) *The Ethics of Research with Children and Young People: A Practical Handbook*, 2nd edn. SAGE Publications Ltd, London, pp. 100–122.

Bailey, S., Boddy, K., Briscoe, S. and Morris, C. (2014) Involving disabled children and young people as partners in research: A systematic review. *Child: Care, Health and Development* 41(4), 505–514.

Blades, R., Renton, Z., Valle, L, I., Clements, K., Gibb, J. *et al.* (2013) We would like to make a change: Children and young people's participation in strategic health decision-making. Available at: https://www.childrenscommissioner.gov.uk/wp-content/uploads/2017/07/FINAL_REPORT_We_would_like_to_make_a_change_Participation_in_health_decisions.pdf (accessed 9 December 2018).

Buhalis, D. and Darcy, S. (2011) *Accessible Tourism: Concepts and Issues*. Channel View Publications, Bristol, UK.

Cameron, L. and Murphy, J. (2006) Obtaining consent to participate in research: The issues involved in including people with a range of learning and communication disabilities. *British Journal of Learning Disabilities* 35(2), 113–120. DOI: 10.1111/j.1468-3156.2006.00404.x.

Canosa, A. and Graham, A. (2016) Ethical tourism research involving children. *Annals of Tourism Research* 61, 219–221. DOI: 10.1016/j.annals.2016.07.006.

Carr, N. (2011) *Children's and Families' Holiday Experiences*. Routledge, New York.

Carter, B. (2009) Tick box for child? The ethical positioning of children as vulnerable, researchers as barbarians and reviewers as overly cautious. *International Journal of Nursing Studies* 46(6), 858–864. DOI: 10.1016/j.ijnurstu.2009.01.003.

Cloquet, I., Palomino, M., Shaw, G., Stephen, G. and Taylor, T. (2018) Disability, social inclusion and the marketing of tourist attractions. *Journal of Sustainable Tourism* 26(2), 221–237. DOI: 10.1080/09669582.2017.1339710.

Conder, J., Schmidt, L. and Mirfin-Veitch, B. (2016) Listening to the voices of children with disabilities in New Zealand. Available at: www.donaldbeasley.org.nz/assets/Voices-of-Children.DBI-Report.pdf (accessed 23 November 2018).

Cooper, H. (2013) The oppressive power of normalcy in the lives of disabled children: Deploying history to denaturalize the notion of the 'normal child'. In: Curran, T. and Runswick-Cole, K. (eds) *Disabled Children's Childhood Studies: Critical Approaches in a Global Context*. Palgrave Macmillan, Basingstoke, UK, pp. 136–151.

Cossar, J., Brandon, M. and Jordan, P. (2016) 'You've got to trust her and she's got to trust you': Children's views on participation in the child protection system. *Child & Family Social Work* 21(1), 103–112. DOI: 10.1111/cfs.12115.

Coyne, I. (2010a) Accessing children as research participants: Examining the role of gatekeepers. *Child: Care, Health and Development* 36(4), 452–454. DOI: 10.1111/j.1365-2214.2009.01012.x.

Coyne, I. (2010b) Research with children and young people: The issue of parental (proxy) consent. *Children & Society* 24(3), 227–237.

Coyne, I. and Carter, B. (2018) Participatory research in the past, present and future. In: Coyne, I. and Carter, B. (eds) *Being Participatory: Researching with Children and Young People: Co-constructing Knowledge Using Creative Techniques*. Springer, Cham, Switzerland, pp. 1–14.
Cullingford, C. (1995) Children's attitudes to holidays overseas. *Tourism Management* 16(2), 121–127. DOI: 10.1016/0261-5177(94)00022-3.
Curran, T. and Runswick-Cole, K. (2014) Disabled children's childhood studies: A distinct approach? *Disability & Society* 29(10), 1617–1630. DOI: 10.1080/09687599.2014.966187.
Darcy, S. (2006) Setting a research agenda for accessible tourism. Sustainable Tourism Cooperative Research Centres (CRC), Australia. Available at: http://citeseerx.ist.psu.edu/viewdoc/download?doi=10.1.1.522.5821&rep=rep1&type=pdf (accessed 23 November 2018).
Davis, J., Watson, N. and Cunningham-Burley, S. (2017) Disabled children, ethnography and unspoken understandings: the Collaborative construction of diverse identities. In: Christensen, P. and James, A. (eds) *Research with Children: Perspectives and Practices*, 3rd edn. Routledge, New York, pp. 121–141.
Figueiredo, E., Eusébio, C. and Kastenholz, E. (2012) How diverse are tourists with disabilities? A pilot study on accessible leisure tourism experiences in Portugal. *International Journal of Tourism Research* 14(6), 531–550. DOI: 10.1002/jtr.1913.
Graburn, N.H.H. (1983) Editor's page. *Annals of Tourism Research* 10(1), 1–3. DOI: 10.1016/0160-7383(83)90045-2.
Graham, A. and Fitzgerald, R. (2010) Children's participation in research: Some possibilities and constraints in the current Australian research environment. *Journal of Sociology* 46(2), 133–147.
Graham, A., Powell, M., Taylor, N., Anderson, D. and Fitzgerald, R. (2013) Ethical research involving children. UNICEF Office of Research–Innocenti, Florence. Available at: https://www.unicef-irc.org/publications/pdf/eric-compendium-approved-digital-web.pdf (accessed 9 December 2018).
Gram, M. (2005) Family holidays. A qualitative analysis of family holiday experiences. *Scandinavian Journal of Hospitality and Tourism* 5(1), 2–22. DOI: 10.1080/15022250510014255.
Groundwater-Smith, S., Dockett, S. and Bottrell, D. (2015) Introduction: Arguing the case for participatory research with children and young people. In: Groundwater-Smith, S., Dockett, S. and Bottrell, D. (eds) *Participatory Research with Children and Young People*. Sage, London, pp. 1–18.
Health and Disability Ethics Committees (2018) Participants with intellectual disabilities – guidance. Available at: https://ethics.health.govt.nz/guides-templates-forms-0/potentially-vulnerable-study-participants-%E2%80%93-guidance/participants (accessed 2 December 2018).
Ho, C.-H. and Peng, H.-H. (2017) Travel motivation for Taiwanese hearing-impaired backpackers. *Asia Pacific Journal of Tourism Research* 22(4), 449–464. DOI: 10.1080/10941665.2016.1276464.
Hodge, N. and Runswick-Cole, K. (2013) 'They never pass me the ball': Exposing ableism through the leisure experiences of disabled children, young people and their families. *Children's Geographies* 11(3), 311–325. DOI: 10.1080/14733285.2013.812275.
Holland, S., Renold, E., Ross, N.J. and Hillman, A. (2010) Power, agency and participatory agendas: A critical exploration of young people's engagement in participative qualitative research. *Childhood* 17(3), 360–375. DOI: 10.1177/0907568210369310.
Horner-Johnson, W. and Bailey, D. (2013) Assessing understanding and obtaining consent from adults with intellectual disabilities for a health promotion study. *Journal of Policy and Practice in Intellectual Disabilities* 10(3), 260–265. DOI: 10.1111/jppi.12048.
Jenkin, E., Wilson, E., Murfitt, K., Clarke, M., Campain, R. *et al.* (2015) Inclusive practice for research with children with disability: A guide. Deakin University, Melbourne. Available at: www.voicesofchildrenwithdisability.com/wp-content/uploads/2015/03/DEA-Inclusive-Practice-Research_ACCESSIBLE.pdf (accessed 2 December 2018).
Kim, S. and Lehto, X.Y. (2013) Travel by families with children possessing disabilities: Motives and activities. *Tourism Management* 37, 13–24. DOI: 10.1016/j.tourman.2012.12.011.
Lansdown, G. (2011) Every child's right to be heard: A resource guide on the UN committee on the rights of the child general comment no. 12. The Save the Children Fund, UK. Available at: https://www.unicef.org/adolescence/files/Every_Childs_Right_to_be_Heard.pdf (accessed 23 November 2018).
Lehto, X., Luo, W., Miao, L. and Ghiselli, R.F. (2018) Shared tourism experience of individuals with disabilities and their caregivers. *Journal of Destination Marketing & Management* 8, 185–193. DOI: 10.1016/j.jdmm.2017.04.001.
Loi, K.I. and Kong, W.H. (2017) Tourism for all: Challenges and issues faced by people with vision impairment. *Tourism Planning & Development* 14(2), 181–197. DOI: 10.1080/21568316.2016.1204357.

Loveridge, J. and Cornforth, S. (2014) The ages of consent: Re-working consensual frameworks in postmodern times. *International Journal of Qualitative Studies in Education* 27(4), 454–471. DOI: 10.1080/09518398.2013.780316.
Mandell, N. (1988) The least-adult role in studying children. *Journal of Contemporary Ethnography* 16(4), 433–467. DOI: 10.1177/0891241688164002.
Mason, J. and Hood, S. (2011) Exploring issues of children as actors in social research. *Children and Youth Services Review* 33(4), 490–495. DOI: 10.1016/j.childyouth.2010.05.011.
Matthews, N. (2009) Contesting representations of disabled children in picture-books: Visibility, the body and the social model of disability. *Children's Geographies* 7(1), 37–49. DOI: 10.1080/14733280802631005.
McCabe, S. (2009) Who needs a holiday? Evaluating social tourism. *Annals of Tourism Research* 36(4), 667–688. DOI: 10.1016/j.annals.2009.06.005.
McKercher, B. and Darcy, S. (2018) Re-conceptualizing barriers to travel by people with disabilities. *Tourism Management Perspectives* 26, 59–66. DOI: 10.1016/j.tmp.2018.01.003.
Messiou, K. (2006) Understanding marginalisation in education: The voice of children. *European Journal of Psychology of Education* 21(3), 305–318. DOI: 10.1007/BF03173418.
Michopoulou, E., Darcy, S., Ambrose, I. and Buhalis, D. (2015) Accessible tourism futures: The world we dream to live in and the opportunities we hope to have. *Journal of Tourism Futures* 1(3), 179–188. DOI: 10.1108/JTF-08-2015-0043.
Minnaert, L., Maitland, R. and Miller, G. (2006) Social tourism and its ethical foundations. *Tourism Culture & Communication* 7(1), 7–17. DOI: 10.3727/109830406778493533.
Moore, T., Saunders, V. and McArthur, M. (2011) Championing choice – lessons learned from children and young people about research and their involvement. *Child Indicators Research* 4(2), 249–267. DOI: 10.1007/s12187-010-9083-1.
Murray, M. and Sproats, J. (1990) The disabled traveller: Tourism and disability in Australia. *Journal of Tourism Studies* 1(1), 9–14.
Nickerson, N.P. and Jurowski, C. (2001) The influence of children on vacation travel patterns. *Journal of Vacation Marketing* 7(1), 19–30. DOI: 10.1177/135676670100700102.
Nyman, E., Westin, K. and Carson, D. (2018) Tourism destination choice sets for families with wheelchair-bound children. *Tourism Recreation Research* 43(1), 26–38. DOI: 10.1080/02508281.2017.1362172.
O'Neill, J. (2014) Voice and the ethics of children's agency in educational research. *New Zealand Journal of Educational Studies* 49(2), 219.
Page, S.J., Innes, A. and Cutler, C. (2015) Developing dementia-friendly tourism destinations: An exploratory analysis. *Journal of Travel Research* 54(4), 467–481.
Phelan, S.K. and Kinsella, E.A. (2013) Picture this safety, dignity, and voice – ethical research with children: practical considerations for the reflexive researcher. *Qualitative Inquiry* 19(2), 81–90.
Pickering, D. (2018) The embodiment of disabled children and young people's voices about participating in recreational activities': Shared perspectives. In: Runswick-Cole, K., Curran, T. and Liddiard, K. (eds) *The Palgrave Handbook of Disabled Children's Childhood Studies*. Palgrave Macmillan, London, pp. 101–124.
Powell, M.A. and Smith, A.B. (2009) Children's participation rights in research. *Childhood* 16(1), 124–142. DOI: 10.1177/0907568208101694.
Prout, A. and James, A. (2015) A new paradigm for the sociology of childhood?: Provenance, promise and problems. In: James, A. and Prout, A. (eds) *Constructing and Reconstructing Childhood: Contemporary Issues in the Sociological Study of Childhood*. Routledge, New York, pp. 6–28.
Pufall, P.B. and Unsworth, R.P. (2004) Introduction: The imperative and the process for rethinking childhood. In: Pufall, P.B. and Unsworth, R.P. (eds) *Rethinking Childhood*. Rutgers University Press, London, pp. 1–21.
Qvortrup, J. (2015) A voice for children in statistical and social accounting: a plea for children's right to be heard. In: James, A. and Prout, A. (eds) *Constructing and Reconstructing Childhood: Contemporary Issues in the Sociological Study of Childhood*. Routledge, New York, pp. 74–93.
Raffety, E.L. (2015) Minimizing social distance: Participatory research with children. *Childhood* 22(3), 409–422. DOI: 10.1177/0907568214548283.
Reiner, R.C., Olsen, H.E., Ikeda, C.T., Echko, M.M., Ballestreros, K.E. *et al*. (2019) Diseases, injuries, and risk factors in child and adolescent health, 1990 to 2017: Findings from the global burden of diseases, injuries, and risk factors 2017 study. *JAMA Pediatrics* 173(6), 1–17.
Richards, V., Pritchard, A. and Morgan, N. (2010) (Re)Envisioning tourism and visual impairment. *Annals of Tourism Research* 37(4), 1097–1116. DOI: 10.1016/j.annals.2010.04.011.

Richards, S., Clark, J. and Boggis, A. (2015) Privileging voices. In: Richards, S., Clark, J. and Boggis, A. (eds) *Ethical Research with Children: Untold Narratives and Taboos*. Palgrave Macmillan, New York, pp. 125–145.

Richards, S. and Clark, J. (2018) Research with disabled children: Tracing the past, present and future. In: Boggis, A. (ed.) *Dis/abled Childhoods? A Transdisciplinary Approach*. Palgrave Macmillan, Cham, Switzerland, pp. 187–209.

Roberts, H. (2017) Listening to children: And hearing them. In: Christensen, P. and James, A. (eds) *Research with Children: Perspectives and Practices*, 3rd edn. Routledge, New York, pp. 142–159.

Runswick-Cole, K., Curran, T. and Liddiard, K. (2018a) Introduction. In: Runswick-Cole, K., Curran, T. and Liddiard, K. (eds) *The Palgrave Handbook of Disabled Children's Childhood Studies*. Palgrave Macmillan, London, pp. xxi–xxiv.

Runswick-Cole, K., Curran, T. and Liddiard, K. (2018b) *The Palgrave Handbook of Disabled Children's Childhood Studies*. Palgrave Macmillan, London.

Sargeant, J. and Gillett-Swan, J.K. (2015) Empowering the disempowered through voice-inclusive practice: Children's views on adult-centric educational provision. *European Educational Research Journal* 14(2), 177–191. DOI: 10.1177/1474904115571800.

Schnoor, O. (2012) Early childhood studies as vocal studies: Examining the social practices of 'giving voice to children's voices' in a crèche. *Childhood* 20(4), 458–471. DOI: 10.1177/0907568212466902.

Schänzel, H.A. (2010) Whole-family research: Towards a methodology in tourism for encompassing generation, gender, and group dynamic perspectives. *Tourism Analysis: An Interdisciplinary Journal* 15(5), 555–569. DOI: 10.3727/108354210X12889831783314.

Schänzel, H.A. and Yeoman, I. (2015) Trends in family tourism. *Journal of Tourism Futures* 1(2), 141–147. DOI: 10.1108/JTF-12-2014-0006.

Sedgley, D., Pritchard, A., Morgan, N. and Hanna, P. (2017) Tourism and autism: Journeys of mixed emotions. *Annals of Tourism Research* 66(14–25), 14–25. DOI: 10.1016/j.annals.2017.05.009.

Shaw, C., Brady, L.M. and Davey, C. (2011) Guidelines for research with children and young people. National Children's Bureau, London. Available at: http://citeseerx.ist.psu.edu/viewdoc/download?doi=10.1.1.370.3966&rep=rep1&type=pdf (accessed 9 December 2018).

Singh, R. and Nayak, J. (2016) Parent-adolescent conflict and choice of conflict resolution strategy: Familial holiday planning. *International Journal of Conflict Management* 27(1), 88–115.

Skyrme, S.L. and Woods, S. (2018) Researching disabled children and young people's views on decision-making: Working reflexively to rethink vulnerability. *Childhood* 25(3), 355–368. DOI: 10.1177/0907568218769365.

Small, J. (2008) The absence of childhood in tourism studies. *Annals of Tourism Research* 35(3), 772–789. DOI: 10.1016/j.annals.2008.06.002.

Smith, R.W. (1987) Leisure of disable tourists. *Annals of Tourism Research* 14(3), 376–389. DOI: 10.1016/0160-7383(87)90109-5.

Spriggs, M. (2010) Understanding consent in research involving children: The ethical issues. Royal Children's Hospital Melbourne, Melbourne. Available at: https://research.vu.edu.au/ordsite/ethics/ethics_childlren_handbook.pdf (accessed 9 December 2018).

Stafford, L. (2017) 'What about my voice': Emancipating the voices of children with disabilities through participant-centred methods. *Children's Geographies* 15(5), 600–613. DOI: 10.1080/14733285.2017.1295134.

Thornton, P.R., Shaw, G. and Williams, A.M. (1997) Tourist group holiday decision-making and behaviour: The influence of children. *Tourism Management* 18(5), 287–297. DOI: 10.1016/S0261-5177(97)00017-4.

Tisdall, E.K.M. (2017) Conceptualising children and young people's participation: Examining vulnerability, social accountability and co-production. *The International Journal of Human Rights* 21(1), 59–75. DOI: 10.1080/13642987.2016.1248125.

UNICEF (2013) Take us seriously! Engaging children with disabilities in decisions affecting their lives. Available at: https://www.unicef.org/disabilities/files/Take_Us_Seriously.pdf (accessed 17 December 2018).

UNICEF (2017) Including children with disabilities in humanitarian action. Available at: http://training.unicef.org/disability/emergencies/index.html (accessed 17 December 2018).

UNICEF (2018) Disabilities: Introduction. Available at: https://www.unicef.org/disabilities/index_65841.html (accessed 17 December 2018).

United Nations (1989) Convention on the rights of the child. Available at: https://www.ohchr.org/en/professionalinterest/pages/crc.aspx (accessed 23 November 2018).

United Nations (2006) Convention on the rights of persons with disabilities – articles. Available at: https://www.un.org/development/desa/disabilities/convention-on-the-rights-of-persons-with-disabilities/convention-on-the-rights-of-persons-with-disabilities-2.html (accessed 23 November 2018).

Watson, N. (2012) Theorising the lives of disabled children: How can disability theory help? *Children & Society* 26(3), 192–202. DOI: 10.1111/j.1099-0860.2012.00432.x.

Wickenden, M. and Kembhavi-Tam, G. (2014) Ask us too! Doing participatory research with disabled children in the global south. *Childhood* 21(3), 400–417. DOI: 10.1177/0907568214525426.

Yau, M.K., McKercher, B. and Packer, T.L. (2004) Traveling with a disability: More than an access issue. *Annals of Tourism Research* 31(4), 946–960.

Zenko, Z. and Sardi, V. (2014) Systemic thinking for socially responsible innovations in social tourism for people with disabilities. *Kybernetes* 43(3/4), 652–666. DOI: 10.1108/K-09-2013-0211.

9 '*The best holidays I ever had*': The Benefits of Social Tourism Programmes for Children at Risk of Poverty and Social Exclusion

Cátia Loureiro[1], Celeste Eusébio[2*] and Elisabete Figueiredo[3]
[1]Department of Economics, Management, Industrial Engineering and Tourism, University of Aveiro, Portugal; [2]Department of Economics, Management, Industrial Engineering and Tourism, University of Aveiro and GOVCOPP Research Unit, Aveiro, Portugal; [3]Department of Social, Political and Territorial Sciences, University of Aveiro and GOVCOPP Research Unit, Aveiro, Portugal

9.1 Introduction

As Small (2008) and Wu *et al.* (2019), among others, claim, children have been absent from tourism research as main subjects of analysis, social tourism studies included. As Feng and Li (2016) recently demonstrated by conducting a systematic literature review, since the end of the 1970s fewer than 50 articles have been published on the topic. In most of the papers, children are considered as passive individuals, subordinated to adults' will (Wu *et al.*, 2019). As Small (2008) and Wu *et al.* (2019) state, the topics researched primarily include the analysis of the role played by children in parental decision-making, children's influence on adults' experiences and, more recently, the travel effects and outcomes for children. This means that, according to Small (2008, p. 773) the majority of tourism research on children has been conducted, with few exceptions, 'from the adult perspective', thus neglecting the children's points of view and their voices. The absence of children tends to be more evident when considering social tourism programmes, frequently designed to promote social inclusion of economically and socially excluded and disadvantaged children.

As in tourism research, and despite the fact that, even in developed countries, poverty and social exclusion tend to affect children's lives in a disproportional manner, children's own perspectives on those processes and dynamics have been underestimated and underresearched (e.g. Attree, 2006; Main, 2019). Recently, however, there is growing interest in considering children's perspectives on the effects of poverty and social exclusion in their lives, specifically on their well-being and health. As Main (2019) states, the inclusion of the children's perspectives in the study of poverty and exclusion has contributed to shedding light on the connection with children's subjective well-being that has been absent from child poverty policies and measures more frequently based on income or on adult perspectives on children's economic

*Corresponding author: celeste.eusebio@ua.pt

DOI: 10.1079/9781789241211.0010

and social deprivation (Crous and Bradshaw, 2017; Main, 2019). As Attree (2006) states, although children have started to be seen as 'experts' regarding their own life and experiences, those from socially excluded groups are still underrepresented. As Attree (2006) also mentions, the long absence of children's views in the poverty and social exclusion literature reflects both the ethical and the methodological difficulties in conducting research with children.

This chapter aims to fill these gaps in the literature regarding both tourism studies and poverty and social exclusion research by adopting a child-centric perspective. Analysing children's participation in the social tourism programme promoted by the 'O Século' Foundation, this chapter aims to unveil children's motivations for participating in the camps, their levels of satisfaction and the perceived benefits. The social tourism programme of the 'O Século' Foundation was founded in 1927 and it mainly consists of organizing holiday camps for economically disadvantaged children in Portugal. During the long life of the programme, several changes have been introduced, accompanying the social and economic changes in Portuguese society and trying to improve children's experiences at the holiday camps. Nowadays, the main ambition of the 'O Século' holiday camp programme is to promote children's social inclusion and well-being. Through the analysis of 131 questionnaires and 132 drawings, collected during the 2014 summer holiday camps of the 'O Século' Foundation, this chapter discusses the benefits of social tourism programmes for children from their own perspectives and experiences.

The chapter is organized into four main sections in addition to this introduction. The following section is devoted to a literature review, discussing both the constraints of poverty and social exclusion in childhood and the relevance of social tourism programmes in mitigating its effects and promoting social inclusion. The methodological design of the research, together with the brief presentation of the social tourism programme promoted by the 'O Século' Foundation, are presented in the third section. Section 9.4 presents and discusses the results in terms of children's motivations, satisfaction levels and benefits perceived. Finally, the main conclusions are condensed in the last section of the chapter.

9.2 Literature Review

9.2.1 Poverty and social exclusion in childhood

Poverty and social exclusion in childhood have been, both jointly and separately, the objects of many studies in different research fields. These notions, although interrelated, are not equivalent. While poverty is mainly related to economic and material deprivation indicators, social exclusion, although including material deprivation, exceeds it, also encompassing feelings of acceptance and belonging to society or a social group (Overmars-Marx *et al.*, 2014). Both concepts are multidimensional and complex, possessing a systemic character whose contours are deepened and rooted in the social fabric. Therefore, it has proven somewhat difficult to define both concepts, mainly due to their broad appropriation and use by public opinion and political discourse. Particularly, as pointed out by Levitas (1998), Overmars-Marx *et al.* (2014) and Gross-Manos (2017), in addition to the difficulty in measuring and in defining social exclusion as a separate notion from poverty – especially referring to childhood – little research has been conducted exclusively on its processes and dynamics. The concept of social exclusion was originally defined by Lenoir (1974) to refer to the 'misfits' and to the most marginalized elements in society. Lenoir (1974) considered that the excluded constituted a social category, a set of individuals connected by common characteristics that reveal their maladaptation to society. He refers to people that lack basic citizenship rights and who had been removed from social protection schemes and social life. As Gross-Manos (2017) states, since Lenoir's proposal of the concept, the term has widened, gaining new and complex definitions and including new social meanings and categories. Social exclusion may, therefore, be defined as the process that leads to the weakening of ties with society, to the production of inequalities and to the lack or denial of access to resources, services and benefits that the majority of the population enjoy, affecting people's place in society, as well as social cohesion and individuals' well-being and quality of life (e.g. Crous and Bradshaw, 2017).

As stated, poverty plays a significant role in social exclusion processes and dynamics and become more central in childhood. As Herranz Aguayo *et al.* (2015) state, children growing up in poor families are at higher risk of social exclusion, the effects of which may continue through adult life. Like the notion of social exclusion, the notion of poverty has also widened over time, as Viñas *et al.* (2019) state. These authors identify four main perspectives regarding the evolution of scientific research on poverty. The first relates to the failure in covering physical human needs. The second refers to basic needs, satisfaction and access to basic services such as health and education. The third is related to the definition of the poverty line and its variation in accordance with the standards of living defined in each context. Finally, the last, and currently widely embraced perspective, has the subjective quality of life and well-being as key components, also following the conclusions of Andresen and Meiland (2017), Gross-Manos (2017) and Main (2019). As Andresen and Meiland (2017) note, especially when referring to child poverty studies, the recognition of the multidimensional character of this notion places the accent not only on material and objective dimensions but also on subjective and less quantifiable aspects, therefore bringing the notion of poverty closer to that of social exclusion.

The multidimensionality, complexity and dynamics of both poverty and social exclusion processes, seem to be particularly relevant when addressing children experiences. As Viñas *et al.* (2019) discussed, these processes and dynamics affect children in a different way from adults, because they have diverse basic needs and they subjectively feel poverty and social exclusion in different ways. Additionally, children seem to have a higher risk of poverty and social exclusion because, as Andresen and Meiland (2017) state, they were born into poverty or poverty simply happens to them. Furthermore children's lack of autonomy, meaning that they are (as Andresen and Meiland, 2017 and Main, 2019 also note) dependent on family and community decisions regarding the distribution of resources. Following the authors' remarks, the vulnerability of children to poverty and social exclusion is generally equally marked by the lack of participation in various social spheres and by the lack of confidence in the future. The findings of the aforementioned authors seem to somehow contradict the assumption that children are primarily focused on the present. Attree (2006), based on a systematic literature review, also states that living in poverty and social exclusion not only damages the present life expectations of children, but also impacts on their aspirations and desires for the future.

The current and increasing child-centric approaches contrast with the (still) dominant methodologies centred on family and households and, according to Main (2019), they have been able to provide direct insights and information regarding the specific resources and perceptions of children. As will be discussed in the following section, social tourism programmes and activities may play a powerful role in stimulating children's well-being, self-esteem, learning and social participation, potentially contributing to mitigating the effects of poverty and promoting social inclusion of disadvantaged children.

9.2.2 Social tourism programmes for economically disadvantaged children

As Minnaert *et al.* (2009) state, although social tourism for economically deprived groups forms part of the social policy in various European countries, little research evidence of its benefits exists, particularly regarding children. Hunziker (1951) first defined social tourism as the tourism activities promoting the participation of economically disadvantaged elements in society. McCabe and Johnson (2013) add that these activities are fostered by financed and social policy measures. According to Minnaert *et al.* (2009), social tourism includes a wide range of commercial and non-commercial, governmental and non-governmental initiatives and activities aiming at offering holiday experiences to people that otherwise would not have them. Examples of this type of tourism include holiday activities for people with disabilities, charity holidays for socially and economically disadvantaged persons (e.g. children, senior citizens) and also the development of community-based tourism initiatives in less favoured areas (Minnaert *et al.*, 2009). Therefore, social tourism creates benefits not only for its participants but also for society as a whole, contributing to enhancing social capital and fostering social inclusion (Sedgley *et al.*, 2012).

Despite the still scarce literature on this field, as previously mentioned, the benefits of participating in social tourism programmes for the most disadvantaged groups of society are widely recognized (Minnaert *et al.*, 2013), particularly regarding children in situations of poverty and social exclusion as the studies of Quinn *et al.* (2008) and McCabe (2009) demonstrate. Recently, besides Minnaert *et al.* (2013), other studies have contributed to broadening the debate around social tourism's effects on, among other, health, well-being and happiness levels (e.g. McCabe and Johnson, 2013; Sedgley *et al.*, 2012).

Among the main benefits of social tourism vacation programmes, especially for children in situations of poverty and social exclusion, and following the conclusions of the abovementioned authors, its contribution has been found to: (i) break routines and relieve the pressures of everyday life; (ii) help participants get to know new places and practise new activities; (iii) increase levels of life satisfaction; (iv) increase social interaction; (v) create and strengthen relationships; (vi) improve self-esteem and confidence levels; (vii) acquire new skills; (viii) improve physical and mental health; (ix) rehabilitate emotional strengths; and (x) experience fun and create happy memories. Quinn *et al.* (2008) mention that the benefits that economically disadvantaged children obtain from participation in leisure, recreation and tourism activities are similar to the benefits obtained by children without financial constraints (Gilbert and Abdullah, 2004; Dolnicar *et al.*, 2012). However, economically disadvantaged children tend to feel those benefits more intensively as tourism activities have a crucial role in improving their overall well-being and sense of happiness, as well as in alleviating the effects of poverty and social exclusion on their lives. Table 9.1 displays the inverse relationship between the benefits identified by Quinn *et al.* (2008) and McCabe (2009) and the effects of poverty and social exclusion on children's lives referred to by Capucha (2010).

Although social tourism programmes have a long tradition in Portugal, the INATEL Foundation (as referred to in the report 'Calypso study compendium of good practices'; Varona, 2010) seems to be the only Portuguese entity internationally recognized in this field. This is due both to the successful implementation

Table 9.1. Relationship between the benefits of tourism and the effects of poverty and social exclusion in children's lives. (Based on Capucha, 2010; Quinn *et al.*, 2008; McCabe, 2009.)

Effects of poverty and social exclusion	Benefits of tourism to children's well-being and social inclusion
• **Daily exposure to stress, violence, criminality, insecurity and to various forms of deprivation**	• Escape from everyday routine and difficult circumstances • Opportunity to have fun and to create happy memories of childhood
• **Emotional instability**	• Rehabilitation of emotional stress
• **Demotivation towards life and their own self**	• Satisfaction with life as a whole
• **Health vulnerability**	• Enhancement of physical and mental health
• **Inability to participate in cultural, recreational and leisure activities**	• Opportunity to know new places and try new activities
• **Undervalued image of oneself** • **Lack of adequate social qualifications and skills**	• Acquisition of new skills • Self-improvement • Improvement of self-esteem and confidence
• **Isolation**	• Improvement of social interaction skills and opportunities
• **Problems in establishing relationships** • **Emotionally unstable relationships**	• Creation and/or strengthening of relationships • Emotional rehabilitation

of several social tourism programmes at the European Union level, and to the fact that the INATEL Foundation is the only Portuguese member of the International Social Tourism Organisation (ISTO). On the one hand, this evidences the excellent work carried out by INATEL, but, on the other hand, this also reveals the lack of knowledge, at least at the international level, of other Portuguese social tourism programmes. Therefore, besides INATEL – which works mainly in the fields of social and senior tourism (Fundação INATEL, 2019) – other organizations in Portugal undertake social tourism programmes providing access to vacations for poor and socially excluded children. Good examples are 'O Século' Foundation – analysed in the chapter presented here – the Foundation Bissaya Barreto (FBB), the 'Associação Para a Animação de Crianças e Jovens' (APACJ) and the 'Banco de Informação Pais a Pais' (BIPP). However, these organizations have a wide range of other social support services and activities, directed at the most vulnerable groups in society. Holiday camps for disadvantaged children are the most common social tourism programmes carried out by these organizations.

According to the 'Conselho Nacional da Juventude' (CNJ, 2009), holiday camps are privileged places for leisure time vacations, where children can share experiences and live in different spaces from the ones they experience in their daily lives. This type of camps provides structured and pedagogical activities, which combine fun and learning, autonomy, responsibility and social interaction, and which allow children to discover their own personal identities and enhance their development at various levels, particularly in sociocultural and educational terms. The main aims of these programmes are in line with the abovementioned benefits.

9.3 Methodology

9.3.1 Brief description of the social programme – holiday camps of the 'O Século' Foundation

Despite the evident role in and contribution to improving economically disadvantaged children's well-being, there is an evident lack of studies about social tourism programmes and organizations in Portugal. As mentioned in the introductory section, the study presented here aims at filling this gap in the literature by analysing the social tourism programme of the 'O Século' Foundation for economically disadvantaged children. This programme, mainly consisting of the organization of holiday camps, started in 1927 and was devised by João Pereira Rosa (director of *O Século* newspaper). The initial aim of the programme was to provide economically disadvantaged children with the opportunity to have a different and rewarding experience from their everyday lives. Since 1927, a significant number of children have benefited from vacations in the holiday camp of 'O Século' Foundation. During the long life of the programme, several changes have been introduced in order to improve children's experiences. Currently the programme aims at giving children the opportunity to experience rewarding vacations in contact with diverse communities and territories and participating in group experiences as a way of fostering social inclusion and integration (Fundação 'O Século', 2019). These experiences will contribute to developing children's creative capabilities, increasing their knowledge and improving their general well-being. In recent years, the programme has promoted 1-week vacations. Participants are different in every vacation period, in order to reach and include a larger number of children. Children may be registered in the programme by their families, sheltering childcare institutions, municipalities and institutions of social solidarity. With a few exceptions (e.g. 13- or 14-year-old children who are at risk of being left alone at home), participants must be between 6 and 12 years old. For 1 week, the children live in the holiday camps and participate in various educational and recreational activities (specially designed for them), monitored 24 h/day by trained monitors. Besides the monitors (generally one person responsible for a group of five children), the team at the holiday camps is composed of a technical director, two coordinators, one medical doctor, one nurse and one psychologist. A great number of the holiday camps take place at the seaside, owing to the privileged location of the Foundation facilities. The number of participants in the holiday camps of the 'O Século' Foundation has increased during recent years.

9.3.2 Data collection methods

To analyse the role of tourism in promoting the well-being and social inclusion of economically disadvantaged children, an empirical study was carried out with a sample of children aged between 6 and 14 years old who participated in the holiday camps offered by the 'O Século' Foundation in 2014. In that year, 214 children participated in the programme, of whom 175 collaborated in the research. Children's perspectives and experiences were analysed, taking into account the motivations for taking part in the holiday camps, the activities carried out during the programme and the perceived benefits.

As previously mentioned, research with children is simultaneously extremely rewarding and complex, raising several ethical and methodological issues. Although children are credible and competent informants of their own needs, activities performed and the perception of benefits, they are, at the same time, generally vulnerable, lacking autonomy and dependent on adult protection (Kirk, 2007; Almeida, 2009). In order to protect the children that participated in the research, and especially considering their age, some precautions were taken while collecting data. First, the children's legal guardians' consent was obtained. Second, the design of the data collection tools was adapted to the children's age. Finally, the children's consent was also obtained and their privacy was guaranteed, together with the confidentiality of the data collected.

The collection of data was based on a combination of quantitative and qualitative methods and techniques, corresponding to questionnaires and drawings. Using questionnaires to obtain information from children frequently implies children's mastery of certain skills, such as the ability to read, to write, interpret and formulate sentences. However, children of 6 or 7 years old generally do not possess those skills and demonstrate difficulties in filling in a questionnaire. In these cases, the use of drawings, increasingly utilized in research with children (Kirk, 2007; Katz and Hamama, 2013), seemed the best way of obtaining information from the 6- and 7-year-old participants. Besides drawing, children between 8 and 14 years old also responded to a questionnaire.

In order for the questionnaire to be adapted to young respondents, it was designed in a simple manner, using simple language and pictures to simultaneously illustrate some questions and make it easy to understand. Although the questionnaire was composed mainly of closed questions, some open questions were used in a complementary manner to provide the researchers with deeper knowledge regarding the children's motivations and their perception of the benefits obtained from their participation in the holiday camps. Open questions also constituted an opportunity for children to express their views and perspectives in a spontaneous manner. The questionnaire included four sections. The first was devoted to the explanation of the research main aims and to the presentation of the instructions on how to fill out the questionnaire. The second section included questions related to the children's leisure and tourism practices. The third section was composed of questions about the children's participation and experience in the holiday camps. Finally, the fourth section included some questions related to the sociodemographic profile of the children surveyed.

To evaluate the leisure and recreation activities in which children participated during the social tourism programme, a list of 14 activities – built upon the studies of Pereira and Neto (1999) and Gram (2005) – was included in the questionnaire. A 5-point Likert-type scale, (in which 1 corresponds to never and 5 to frequently) was used to measure children's participation in each activity. The motivation for participating in the programme was evaluated using eight items (following the studies of Larsen and Jenssen, 2004; Hilbrecht *et al.*, 2008; Quinn *et al.*, 2008), again measured through a 5-point Likert scale, ranging from 1 (strongly disagree) to 5 (strongly agree). Satisfaction with the social tourism programme was evaluated based on six items (based on Quinn *et al.*, 2008) and the satisfaction with the leisure and recreation activities was evaluated using a list of 12 leisure and recreation activities. Both satisfaction with the programme and satisfaction with the activities carried out were examined using again a 5-point Likert-type scale, ranging from 1 (I did not like it) to 5 (I liked it very much). Finally, the benefits perceived by the children from their participation in the programme were evaluated using a list of 12 benefits, (based on the studies of Quinn *et al.*, 2008 and McCabe, 2009) measured again through a 5-point Likert scale varying from 1 (strongly disagree) to 5 (strongly agree). A pilot test with 15 children aged between 8 and 12 years old was carried out in May 2014, to examine the adequacy, the readability and possible ambiguity of the questionnaire. Based on the pilot test results, minor adjustments were introduced in the questionnaire.

As mentioned, drawing has been recognized as one of the most important tools to evaluate feelings, emotions and personality traits of children (Malchiodi, 1998; Katz and Hamama, 2013). Through drawing, children can manifest certain feelings, ideas and opinions that they are not easily able to express through oral or written communication. In this research, all the children participating in the social tourism programme were invited to draw on a blank sheet with the following sentence on top: 'Use this sheet to draw your vacation this year at the holiday camp of the "O Século" Foundation.' As explained before, drawing was the only tool used to collect information from the 6- and 7-year-old participants. In the case of children between 8 and 14 years old, drawing, although not used by all, was also extremely useful to complete the information obtained through the questionnaire.

During 2014, three holiday camps were organized by the 'O Século' Foundation. The first one from 23 June to 4 July, the second one from 7–18 July and the third one from 21 July to 1 August. The collection of data, both through questionnaires and drawings, took place on the penultimate day of the holiday period in each of the three holiday camps, always after lunch and before the beginning of the afternoon activities (between 2:00pm and 3:00pm) in order to interfere as little as possible in the programme. A total of 131 questionnaires and 132 drawings were collected. Of the drawings 116 were black and white and 16 were coloured.

9.3.3 Data analysis methods

Data obtained from the questionnaire were analysed using SPSS Version 21. Descriptive statistical analysis were used to characterize the sample, children's participation in leisure and recreation activities, their motivation for participating in this social programme, the activities programmed, children's satisfaction with the overall programme and with the different activities as well as their perception of the benefits obtained. Principal component analysis (PCA) was performed both to identify dimensions of the motivations to participate in the programme and the perception of the benefits obtained. The content of drawings was analysed through the examination of the elements most represented and of the colours used.

9.4 Results and Discussion

9.4.1 Sample profile

Regarding the questionnaire respondents, children aged 12 years old (26.00%) predominate in the sample, followed by 11- and 9-year-olds with the same percentage (17.60%). Eight-year-olds (13.00%) and children between 13 and 14 years old (11.40%) are the smallest groups in the sample. The majority of the children are female (55.70%).

Regarding children's tourism practices during their vacations, about 84% usually take holidays away from home, revealing a low level of deprivation of the sample in terms of tourism participation. However, about 41% of the children usually spend their vacations in holiday camps promoted by social solidarity institutions. This finding highlights the importance of these types of social initiatives in providing tourism experiences to economically disadvantaged children. As Minnaert *et al.* (2011) mention, social tourism is often the only opportunity to fulfil the right to tourism for the most disadvantaged social groups, children included.

Regarding the type of activities performed by children during their holidays, the findings presented in Table 9.2 reveal that the most frequent activities performed are: playing with friends, watching TV and going to the beach. Going to a water park, zoo or aquarium, camping, visiting amusement parks, museums and monuments are the least performed activities. These findings clearly demonstrate the importance of non-paid activities in the children's holidays, which may indicate not only economic

Table 9.2. Activities performed by the children during their vacations (in general).

Activities	*N*	Frequency (%) 1 – Never	Frequency (%) 5 – Frequently	Mean	Standard deviation
Playing with friends	131	1.50%	60.30%	4.27	1.031
Watching TV	130	9.20%	43.80%	3.96	1.241
Going to the beach	131	3.10%	33.60%	3.84	1.066
Going to playgrounds	131	6.10%	19.10%	3.33	1.153
Computer games	131	17.60%	24.40%	3.30	1.418
Cycling	129	17.80%	23.30%	3.29	1.400
Going to the swimming pool	131	3.80%	12.20%	3.21	0.985
Internet	131	26.00%	26.70%	3.14	1.543
Reading	131	13.70%	17.60%	3.11	1.269
Visiting museums and monuments	131	32.10%	1.50%	2.10	0.991
Going to amusement parks	131	47.30%	6.90%	2.07	1.260
Camping	131	50.40%	3.10%	1.98	1.183
Going to zoos or aquariums	129	36.40%	0.80%	1.95	0.926
Going to water parks	130	55.40%	4.60%	1.82	1.138

constraints but also the lack of opportunity to participate in paid activities that often leads to aggravate children's feelings of social exclusion. Therefore, it seems of utmost importance that these children could participate in the same type of activities as the non-economically deprived children such as going to the zoo, visiting water parks, museums and monuments.

9.4.2 Motivations for participating in the social tourism programme

The PCA results (Table 9.3) reveal that children were highly motivated to participate in the social tourism programmes. The two factors generated from the PCA (F1: *Freedom and adventure* and F2: *Socialization*) demonstrate that children participate in the programme mainly to make new friends, have fun, play with other children and experience new adventures. The children surveyed are, however, less motivated to participate in the holiday camps to win the respect of others, feel free or rest from their daily routines (mean below 4.30).

The importance of creating new friendships was also highlighted in the works of Larsen and Jenssen (2004) and Quinn *et al.* (2008). Larsen and Jenssen (2004) observed that the most important motivations for children's participation in study visits are not the place visited, nor the opportunity to learn, but rather travelling with friends. The authors associate this motivation with the fact that children view these visits as an opportunity to develop their sense of belonging, to test their own identity and to build social bonds. Quinn *et al.* (2008) also found that for children with previous experience in social tourism programmes, the possibility of encountering friends again and the creation of new friendships are the most important motivations for re-joining a social tourism programme. The importance of creating friendships may be related to the fact that economically disadvantaged children have a limited network of contacts and social participation opportunities.

9.4.3 Perceived benefits from the programme

Children perceived the benefits obtained from their participation in holiday camps as extremely significant. On a scale from 1 (strongly disagree) to 5 (strongly agree), all the benefits presented in Table 9.4 obtained a mean higher

Table 9.3. PCA of motivations for participating in the social tourism programme.

	Mean	Communality	Factor loading	Eigenvalue	Cumulative variance explained (%)	Cronbach's alpha
F1: Freedom and adventure	4.32			2.225	43.023	0.697
Have new adventures	4.44	0.659	0.806			
Learn new things	4.33	0.605	0.749			
Feel free	4.20	0.550	0.711			
F2: Socialization	4.45			2.219	55.550	0.697
Make new friends	4.74	0.612	0.759			
Rest from daily routine	4.21	0.438	0.661			
Playing with other children	4.47	0.596	0.645			
To win the respect and admiration of others	4.12	0.468	0.643			
Have fun	4.70	0.518	0.527			
N = 131	KMO = 0.807		Bartlett's test of sphericity (*P*= 0.000)			

Table 9.4. PCA of the perceived benefits obtained with the participation in the social tourism programme.

Benefits obtained	Mean	Communality	Factor loading	Eigenvalue	Cumulative variance explained (%)	Cronbach's alpha
F1: Friendship and fun	4.60			3.209	48.593	0.804
Make new friends	4.72	0.679	0.809			
Meet new people	4.73	0.593	0.808			
Carry out new activities	4.63	0.546	0.609			
Increase my knowledge	4.31	0.423	0.603			
Fun and create happy memories	4.68	0.434	0.555			
Discovery of new capabilities	4.53	0.540	0.533			
F2: Self-esteem and personal development	4.48			3.193	58.198	0.832
Feel more confident	4.36	0.760	0.867			
Feeling happier	4.47	0.551	0.708			
Get closer to my friends	4.38	0.531	0.667			
Meet new places	4.58	0.678	0.663			
Learn new things	4.60	0.667	0.626			
N = 131	KMO = 0.875		Bartlett's test of sphericity (*P* = 0.000)			

than 4. Therefore, participation in these holiday camps permitted children to have fun and discover new things, as well as promoting their self-esteem and personal development. However, the opportunity to meet new people (mean = 4.73), to make new friends (mean = 4.60), and to create happy and rewarding memories (mean = 4.68) emerge as the most important benefits perceived.

The PCA of the benefits perceived from participation in the social tourism programme accomplishes all the requirements for this type of analysis. Two main factors were found. Factor 1: *friendship and fun* explains 48.59% of the variance of the data and has a high internal consistency (Cronbach's alpha of 0.804), representing the most important dimension of the benefits mentioned by the children (mean = 4.60). Factor 2: *self-esteem and personal development* explains 9.6% of the variance of the data and also has a high internal consistency (Cronbach's alpha of 0.832). This factor has a mean of 4.48, revealing that these kinds of social programmes have great potential for fostering children's self-esteem, as well as promoting their personal development, therefore contributing to the children's learning of new things, getting to know new places and feeling happier.

The children also expressed their perceptions of the holiday camp benefits by answering an open question. Results are in line with the analysis presented in Table 9.4. Children's descriptions highly emphasize the fun aspects: 'I had a lot of fun in the holiday camps'; 'It was the most fun summer'; and 'I described it as a fun and lively vacation'. These results are in line with the conclusions of Hilbrecht *et al.* (2008), where fun also emerged as one of the most important aspects of the family vacation experiences.

Children participating in the 'O Século' holiday camps described additional benefits from the programme, such as the opportunity to learn new things, meet new people and experience several activities. Furthermore, children described the programme as generating happy memories and positive emotions: 'Here I feel good and I learned to have more respect'; 'I think this holiday camp helps us learn many new and funny things'; 'I learn and I help a lot'; 'This year in this holiday camp I had a lot of fun and I am sure that for the next few years I will never forget it'; 'I feel happy to be here'; and 'I had a lot of fun and a lot of emotion'.

9.4.4 Satisfaction and loyalty with the programme

The analysis of the children's satisfaction levels regarding the holiday camps was based on two approaches. First, a global evaluation of the programme was carried out through the analysis of the children's satisfaction level with the activities included in the programme, as well as with the monitors. The data presented in Table 9.5 show that children were extremely satisfied with both dimensions.

The evaluation of the children's satisfaction with the different activities of the programme (Table 9.6) also reveals that children are very satisfied (all means are higher than 4). However, disco, treasure hunt game, century festival, cooking, walks and going to the beach stand out as the most satisfying activities (mean higher than 4.5). On the other hand, karaoke and paper chase are the least satisfying activities.

Children were also invited to express their feelings regarding the experience at the holiday camps by answering an open question. The overall satisfaction with the programme is visible in all the positive descriptions made by the children, such as: 'My holiday are going wonderfully'; 'Fantastic'; 'Very good'; 'Spectacular'; and

Table 9.5. Satisfaction with the dimensions of social tourism programme – holiday camps.

Dimensions of the programme	*N*	Frequency (%)		Mean	Standard deviation
		1 – I did not like it	5 – I liked it very much		
Activities	131	0.00	67.20	4.62	0.588
Human resources (monitors)	131	0.00	74.00	4.66	0.628

Table 9.6. Satisfaction with the activities carried out in social tourism programme – holiday camps.

Activities	N	Frequency (%) 1 – I did not like it	5 – I liked it very much	Mean	Standard deviation
Disco	131	0.00	85.5	4.81	0.528
Treasure hunt game	131	2.30	78.6	4.69	0.755
Festival (Century festival)	129	0.00	76.7	4.69	0.635
Cooking	130	0.80	70.8	4.66	0.604
Walks	131	1.50	64.1	4.53	0.768
Beach	131	0.00	61.8	4.51	0.706
'Mangueirada'	131	1.50	54.2	4.37	0.853
Gymkhana	130	3.80	56.9	4.35	0.963
Ateliers	131	1.50	57.3	4.34	0.917
Paper chase	128	3.10	52.3	4.21	0.993
Karaoke	129	5.40	59.7	4.14	1.210

'I love being here'; and 'The best holidays I ever had'.

Satisfaction is both related to the activities undertaken and to the relationship with monitors. Some children mentioned that: 'I enjoyed going to the beach and the swimming pool'; 'I went to the beach, to the pool and I had a lot of fun at the disco'; 'I really liked the disco'; 'My monitor is very cool'; and 'I love this institution ['O Século'], this holiday camp is my favourite and I really like the monitors'.

9.4.5 Characterization of the holiday camp through children's eyes: Analysis of their drawings

Taking into consideration that through drawing, children express their ideas and perceptions about the world as well as their emotions (Katz and Hamama, 2013), in this research, drawings were used to analyse children's views of their experiences, mainly in terms of benefits obtained, in the holiday camps. A total of 132 drawings were obtained from children between 6 and 14 years old. The analysis of the drawings was based on the identification of the elements represented and on the colours used. Regarding the elements represented in the drawings (Table 9.7), 78% refer to people and 51.5% are related to the leisure and recreation activities carried out during the holiday camps. Eleven per cent also included written messages.

People represented in the drawings are mainly the children (the author of the drawing and/or colleagues and friends) (45.5%) and the monitors (23.5%). The focus on representing colleagues and friends of the author, that is, other children, frequently in situations of interaction, is in line with the data obtained from the questionnaire, reinforcing the relevance of the programme to the participants in terms of socialization and creation of new friendships. As discussed before, the opportunities for interaction with other children are of utmost importance for the participants, who are, in the majority of the cases, in situations of poverty and social exclusion and deprived of these opportunities in their daily lives. These results are in line with the benefits identified by Quinn *et al.* (2008), McCabe (2009), McCabe and Johnson (2013) and Sedgley *et al.* (2012) related to the

Table 9.7. Elements identified in the drawings.

Elements	*N*	%
People	103	78.00
Activities	68	51.50
Natural elements	57	43.20
Facilities and equipment	31	23.50
Written messages	15	11.40

Fig. 9.1. Examples of representations of people in children's drawings. (Drawings by the children interviewees, used with permission.)

contribution of tourism to increase and improve social interaction skills and situations, as well as to create and strengthen relationships.

In the same vein, children represented in the drawings are often engaged in leisure and recreation activities (e.g. playing together on the beach, dancing at the disco), reinforcing the relevance of the participation in the holiday camps for social interaction and creating friendship connections. As mentioned, the monitors are also represented in the drawings in special and careful ways: standing alone with just his/her name; together with the author of the drawing; and/or represented with a considerably larger size compared with the child (Fig. 9.1). These highlight the importance of monitors in the children's experience of the holiday camps, simultaneously suggesting the development of (real or idealized) affective bonds between the children and the monitors or the desire to be protected. This is in accordance with the conclusions of Quinn *et al.* (2008) regarding the role of monitors during social tourism programme experiences. On the one hand, monitors may be seen by the children as their protectors and as responsible for their safety and fun activities during their vacations. On the other hand, monitors may be viewed as someone that is filling a gap in children's lives regarding the absence of one or of both parents.

Leisure and recreation activities were the second more represented features in children's drawings (Fig. 9.2). Going to the beach, 'mangueirada' (using a garden hose to shower the colleagues) and going to the disco were the most frequently represented activities, again in line with the results obtained from the questionnaire in terms of the most satisfying activities performed during the holiday camps and with the benefits perceived as previously discussed. These activities contributed to an experience of fun and escape from daily routines, at the same time contributing to the creation of happy memories regarding the vacation.

Some children also included written messages in their drawings, as a way of better expressing and/or reinforcing their feelings and emotions regarding their experience in the holiday camps. The written messages only express positive feelings and emotions, again corroborating the results obtained from questionnaires. Written messages are usually presented in the drawings in dialogue boxes or, as titles of the drawing, ranging from complimenting the monitors ('Ju, you are the greatest monitor'), feelings towards the 'O Século' Foundation ('I love the Século'), benefits achieved owing to participation in the programme ('Love, Fondness, Happiness, Listening, Friendship, Patience, Learning. These were what I learned to do and to give') and positive descriptions of vacations experienced ('Very fun and cool vacations').

Finally, the colours used in the drawings were analysed in order to examine children's

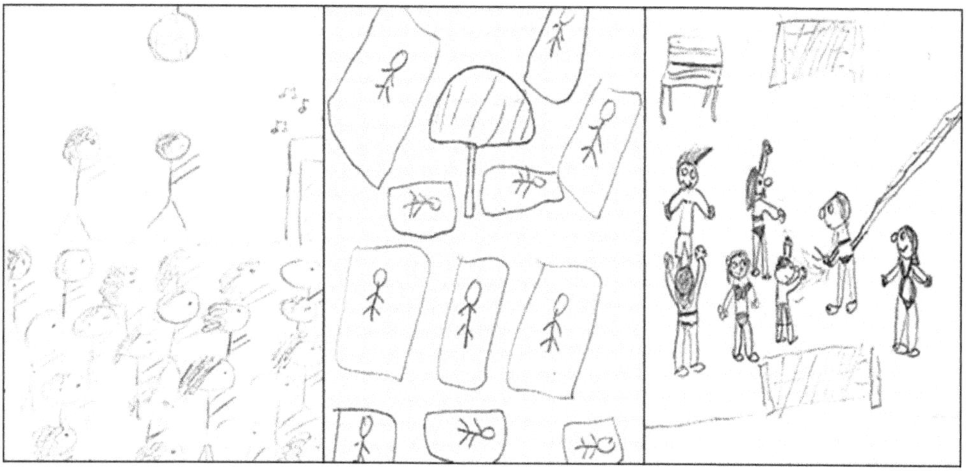

Fig. 9.2. Examples of the leisure and recreation activities represented in children's drawings. (Drawings by the children interviewees, used with permission.)

emotional state (positive or negative) regarding their experiences in the holiday camps. The analysis was based on the study conducted by Boyatzis and Varghese (1994), according to whom children usually relate the brightest colours (e.g. pink, yellow, red, green, purple and blue) to positive emotions (e.g. happiness, excitement and inner strength) and darker colours (e.g. brown, black and grey) to negative emotions (e.g. sadness, anger and annoyance). Despite that, authors also emphasize that boys tend to associate fewer negative feelings and emotions with darker colours when compared with girls. Boyatzis and Varghese (1994), contrary to what other studies suggested, also state that in their research, the colour red was generally related to positive emotions such as happiness and excitement.

Although, as suggested, emotional associations with colours may vary according to certain factors such as gender, age, context, children's cultural characteristics and cognitive development, the colours referred to in the study of Boyatzis and Varghese (1994) were used to analyse the 16 coloured drawings collected from the participants in the 'O Século' holiday camps (Fig. 9.3). Table 9.8 displays the results of that analysis, revealing the predominance of brighter colours, mainly red, yellow and blue.

As demonstrated in the study conducted by Boyatzis and Varghese (1994), although these three colours are mainly related to happiness, red and yellow are also associated with excitement and blue is connected to inner strength. These results are again in accordance with previous studies about the contribution of social tourism programmes to the emotional well-being of children, contributing to rehabilitating their emotional strengths (e.g. Quinn *et al.*, 2008).

Regarding the use of darker colours, the children often utilize brown in their drawings. However, as the study by Boyatzis and Varghese (1994) again demonstrates, of the darker colours, brown is the least related to negative emotions. Examining the 16 colour drawings collected, it became evident that the use of the colour brown is related to the representation of elements that are, in reality, brown, such as tree trunks, soil and hair and not to negative feelings and emotions.

9.5 Conclusions

This chapter aimed to discuss the benefits of social tourism programmes for economically disadvantaged children, namely their contribution to the promotion of well-being, happiness and social inclusion. Aiming at filling the gap in tourism studies and in poverty and social exclusion research regarding children's own perspectives, voices and experiences, this chapter

Fig. 9.3. Examples of the use of colour in children's drawings. (Drawings by the children interviewees, used with permission.)

adopted a child-centric approach by giving voice to economically disadvantaged children who participated in the 2014 holiday camps promoted by the 'O Século' Foundation, in terms of their motivation, satisfaction levels and benefits perceived.

The chapter departs from a theoretical discussion on the meanings and effects of poverty and social exclusion on children's lives and well-being, taking into account that these processes and dynamics disproportionally affect children who, in turn, have fewer resources and skills to deal with their consequences. Poverty and social exclusion in childhood often mean a lack of opportunities to participate in social spheres, a lack of confidence in the future, as well as a lack of

Table 9.8. Colours used in children's drawings. (Based on Boyatzis and Varghese, 1994.)

Emotions	Colour	Number of drawings in which colour appears
Happiness Excitement Inner strength	Rose	9
	Red	12
	Yellow	11
	Green	7
	Purple	3
	Blue	12
Sadness Anger Annoyance	Brown	10
	Black	1
	Grey	0

several types of resources and a lack of access to a wide range of activities and services. Social tourism programmes may alleviate the negative consequences of poverty and social exclusion in children's lives, contributing to promoting social inclusion, namely by alleviating the pressures of everyday life, providing contact with new and different people and places, and allowing access to leisure and recreation activities. At the same time, social tourism programmes contribute to creating and strengthening friendships, improving children's confidence and self-esteem, enabling them to experience fun and rewarding activities and enhancing overall well-being and happiness. Although these may be identified as general positive benefits deriving from tourism activities, economically disadvantaged children tend to experience them more intensively.

Empirical evidence collected through questionnaires and drawings from children aged between 6 and 14 years old that participated in the holiday camps of the 'O Século' Foundation, clearly demonstrated the abovementioned benefits, in line with the conclusions of Quinn *et al.* (2008) and McCabe (2009). Results also reveal that the perceived benefits from participating in the holiday camps – for example, well-being, happiness, satisfaction with social interactions and activities – are in line with the conclusions of McCabe and Johnson (2013) and Sedgley *et al.* (2012). Although 84% of the children usually take holidays away from home, 41% spend their vacation in holiday camps promoted by social solidarity institutions and undertake activities different from their everyday lives, which reveals the significance of social tourism experiences to children at risk of poverty and social exclusion.

Children's overall appreciation of the experience at the holiday camps is extremely positive. The activities performed, the motivations for participating and the benefits perceived strongly highlight the importance of creating new friendship relationships and the opportunities to socialize, following again the conclusions of Quinn *et al.* (2008) and Larsen and Jenssen (2004). Another important benefit perceived relates to experiencing fun and happiness during the social tourism programme of the 'O Século' Foundation, in line with the results highlighted by the study of Hilbrecht *et al.* (2008). The importance of fun and happiness, as well as friendships and social interaction, is also corroborated by the type of activities seen by children as providing higher levels of satisfaction. Children tend to demonstrate higher levels of satisfaction with collective and fun activities, even though they classify all the activities as rewarding.

The elements and colours used in the drawings also emphasize the abovementioned dimensions and benefits, and confirm the overall positive feelings and emotions towards the holiday camp experiences. People (particularly friends and colleagues, the author of the drawing and monitors) and leisure and recreation activities assume a central place in the drawings collected and analysed, which again emphasizes the relevance of social interaction opportunities for the children. In the 16 coloured drawings collected, the predominant use of brighter colours (such as pink, blue, red and yellow) reinforces the positive emotions, as well as feelings of well-being and happiness (in line with the conclusions of Boyatzis and Varghese, 1994) triggered by the children's participation in the social tourism programme.

Although results cannot be generalized – mainly owing to the small size of the sample and to the limited scope of the programme analysed – this research contributes to shedding some light on the contributions of social tourism programmes to economically disadvantaged children's well-being and happiness, from a child-centric perspective.

Acknowledgements

The authors express their gratitude to the 'O Século' Foundation, to the children's parents and legal guardians, and, of course, to all the children that participated in the research.

References

Almeida, A.N. (2009) *Para uma sociologia da infância – Jogos de olhares, pistas para a investigação. [For a sociology of childhood – Games of looks, clues for investigation]*. ICS, Lisbon.

Andresen, S. and Meiland, S. (2017) Being poor from children's point of view. The potential of childhood theory and qualitative methods for child poverty research: Findings from two qualitative studies of poverty in Germany. *Children and Youth Services Review* 97, 94–99. DOI: 10.1016/j.childyouth.2017.07.010.

Attree, P. (2006) The social costs of child poverty: A systematic review of the qualitative evidence. *Children & Society* 20, 54–66.

Boyatzis, C.J. and Varghese, R. (1994) Children's emotional associations. *The Journal of Genetic Psycology* 1, 77–85.

Capucha, L. (2010) Inovação e Justiça Social – Políticas activas para a inclusão educativa. [Innovation and social justice – Active policies for educational inclusion]. *Sociologia, Problemas e Práticas* 63, 25–50.

CNJ (2009) *Campos de Férias em Portugal. [Vacation camps in Portugal]*. CNJ, Lisbon.

Crous, G. and Bradshaw, J. (2017) Child social exclusion. *Children and Youth Services Review* 80, 129–139. DOI: 10.1016/j.childyouth.2017.06.062.

Dolnicar, S., Yanamandram, V. and Cliff, K. (2012) The contribution of vacations to quality of life. *Annals of Tourism Research* 39(1), 59–83. DOI: 10.1016/j.annals.2011.04.015.

Feng, X. and Li, M. (2016) Children tourism: A literature review. *Tourism Tribune* 31(9), 61–71.

Fundação INATEL (2019) Homepage. Available at: www.inatel.pt (accessed 5 March 2019).

Fundação 'O Século' (2019) Homepage. Available at: www.oseculo.pt (accessed 5 March 2019).

Gilbert, D. and Abdullah, J. (2004) Holiday taking and the sense of well-being. *Annals of Tourism Research* 31(1), 103–121. DOI: 10.1016/j.annals.2003.06.001.

Gram, M. (2005) Family holidays. A qualitative analysis of family holiday experiences. *Scandinavian Journal of Hospitality and Tourism* 5(1), 2–22. DOI: 10.1080/15022250510014255.

Gross-Manos, D. (2017) Material well-being and social exclusion association with children's subjective well-being: Cross-national analysis of 14 countries. *Children and Youth Services Review* 80, 116–128. DOI: 10.1016/j.childyouth.2017.06.048.

Herranz Aguayo, I., Díaz Herráiz, E., Montenegro Marques, E., Machado, I. and Almeida, S. (2015) Child at risk of poverty or social exclusion: comparative view between Spain and Portugal in the European context. *Social Indicators Research* 129(3), 961–978. DOI: 10.1007/s11205-015-1119-y.

Hilbrecht, M., Shaw, S.M., Delamere, F.M. and Havitz, M.E. (2008) Experiences, perspectives, and meanings of family vacations for children. *Leisure/Loisir* 32(2), 541–571. DOI: 10.1080/14927713.2008.9651421.

Hunziker, W. (1951) *Social Tourism: Its Nature and Problems*. International Tourists Alliance Scientific Commission, Geneva, Switzerland.

Katz, C. and Hamama, L. (2013) 'Draw me everything that happened to you': Exploring children's drawings of sexual abuse. *Children and Youth Services Review* 35(5), 877–882. DOI: 10.1016/j.childyouth.2013.02.007.

Kirk, S. (2007) Methodological and ethical issues in conducting qualitative research with children and young people: A literature review. *International Journal of Nursing Studies* 44(7), 1250–1260. DOI: 10.1016/j.ijnurstu.2006.08.015.

Larsen, S. and Jenssen, D. (2004) The school trip: Travelling with, not to or from. *Scandinavian Journal of Hospitality and Tourism* 4(1), 44–56. DOI: 10.1080/15022250410006273.

Lenoir, R. (1974) *Les Exclus. Un Français sur dix. [The excluded. One French person out of ten]*. Seuil, Paris.

Levitas, R. (1998) *The Inclusive Society? Social Exclusion and New Labour*. Macmillan Press, London.

Main, G. (2019) Child poverty and subjective well-being: The impact of children's perceptions of fairness and involvement in intra-household sharing. *Children and Youth Services Review* 97, 49–58. DOI: 10.1016/j.childyouth.2017.06.031.
Malchiodi, C.A. (1998) *Understanding Children's Drawings*. The Guilford Press, New York.
McCabe, S. (2009) Who needs a holiday? Evaluating social tourism. *Annals of Tourism Research* 36(4), 667–688. DOI: 10.1016/j.annals.2009.06.005.
McCabe, S. and Johnson, S. (2013) The happiness factor in tourism: subjective well-being and social tourism. *Annals of Tourism Research* 41, 42–65. DOI: 10.1016/j.annals.2012.12.001.
Minnaert, L., Maitland, R. and Miller, G. (2009) Tourism and social policy – the value of social tourism. *Annals of Tourism Research* 36(2), 316–334.
Minnaert, L., Maitland, R. and Miller, G. (2011) What is social tourism? *Current Issues in Tourism* 5(5), 403–415. DOI: 10.1080/13683500.2011.568051.
Minnaert, L., Maitland, R. and Miller, G. (2013) *Social Tourism: Perspectives and Potential*. Routledge, London.
Overmars-Marx, T., Thomése, F., Verdonschot, M. and Meininger, H. (2014) Advancing social inclusion in the neighbourhood for people with an intellectual disability: an exploration of the literature. *Disability & Society* 29(2), 255–274. DOI: 10.1080/09687599.2013.800469.
Pereira, B.O. and Neto, C. (1999) As crianças, o lazer e os tempos livres. [Children, leisure and free time]. In: Sarmento, M.J. (ed.) *Para uma Bibliografia sobre a Infância e as Crianças em Portugal. [For a bibliography on childhood and children in Portugal]. Centro de Estudos da Criança*. Universidade do Minho, Braga, Portugal, pp. 85–107.
Quinn, B., Griffin, K. and Stacey, G. (2008) *Poverty, Social Exclusion and Holidaying: Towards Developing Policy in Ireland*. Combat Poverty Agency, Working Paper.
Sedgley, D., Pritchard, A. and Morgan, N. (2012) 'Tourism poverty' in affluent societies: Voices from inner-city London. *Tourism Management* 33(4), 951–960. DOI: 10.1016/j.tourman.2011.10.001.
Small, J. (2008) The absence of childhood in tourism studies. *Annals of Tourism Research* 35(3), 772–789. DOI: 10.1016/j.annals.2008.06.002.
Varona, H. (2010) *Calypso Study on Social Tourism Portugal*. CCE.
Viñas, F., Casas, F., Abreu, D.P., Alcantara, S.C. and Montserrat, C. (2019) Social disadvantage, subjective well-being and coping strategies in childhood: The case of north-eastern Brazil. *Children and Youth Services Review* 97, 14–21. DOI: 10.1016/j.childyouth.2017.06.012.
Wu, M.-Y., Wall, G., Zu, Y. and Ying, T. (2019) Chinese children's family tourism experiences. *Tourism Management Perspectives* 29, 166–175. DOI: 10.1016/j.tmp.2018.11.003.

10 The Financial Accessibility of Children's Camps: A French Challenge to Renew Their Popularity

Xavier Michel*
University of Caen-Normandy, Caen, France

10.1 Introduction

In France, the participation of children in summer camps has been decreasing for several decades. This decline is due to other holiday trends, security reasons, and the loss of perception of the camps' objectives. Yet one of the main factors in the crisis is the increase in prices for attending a camp. This has resulted from the request for and promotion of themed and costlier activities, more expensive transport and accommodation, and better remuneration for camp leaders. A de-differentiation has been occurring between social tourism organizers and for-profit organizers. As a result, a number of camps cost around €500/week, sometimes more, even though they are operated by a not-for-profit organization. While low-income family children and children in care can benefit from social funding to attend a camp, the issue is of more concern to middle-class families, who receive less financial support or none at all.

Most organizers are convinced that they are unable to reduce their prices notably, and believe that they must follow the trends of costly themed activities and comfortable stays. This trend questions the meaning of the social dimension of these holidays for children, and the children's ability to participate in a camp. Yet other organizers have decided to conceive low-cost camps. Our research aimed to identify them, to grasp their particular ways of offering low-cost camps and to reveal their motives. The first part of the chapter explores the links between social tourism and children's camps in France, assuming that the current camps must be analysed against the models of social tourism. The second part of the chapter specifies the financial issues and outlines the camps' characteristics, giving the price levels.

10.2 Social Tourism and Children's Camps in France

Within the typology of social tourism, children's camps have been described as holidays for 'minors outside of their familial milieu' (Chauvin, 2002, p. 83). Caire (2006, p. 32) reminds us that holiday centres for children and adolescents belong to social tourism, and includes the summer camps within the history of 'the origins of social tourism in France' (Caire, 2012, p. 76). On the other hand, tourism is often a rejected notion both by not-for-profit organizers, and by some researchers (Bocquet and Dheilly, 2018; Chaudière and

*xavier.michel@unicaen.fr

 Social Tourism: Global Challenges and Approaches (eds J. Lima and C. Eusébio)
DOI: 10.1079/9781789241211.0011

Baldacchino, 2018) who represent tourism as just a profitable business, neglecting its social and environmental dimensions. These authors do not consider the notions of educational tourism (Ritchie, 2003; Choudhary, 2010) or slow tourism (Fullagar *et al.*, 2012), which provide a meaning of tourism as a beneficial human activity to discover, encounter and learn, provided that these educational and slow approaches are adopted. More generally, the lack of analysis of childhood in tourism studies (Small, 2008; Obrador, 2012) also resulted in a lack of recognition of children's camps as a form of tourism: tourism and social tourism should be expressions used only for adult holidays. By representing tourism as a whole system that contains beneficial human engagements through the discovery of otherness in a leisure dimension, away from home, for all – which is criticized as 'mass tourism' – we can highlight the similarities between tourism and children's camps in social tourism. However, this thought leads us to another possible explanation for the rejection of children's camps as tourism. There are some who would not have the discovery of otherness as an objective, but would just aim to implement a social life between peers and with their camp leaders without seeking to spread out of the camp; this position could be explained by the tradition of a school form in camps (Houssaye, 1998). In this enclaved kind of space, the camps would not be objectively a form of tourism.

Thus, do children's camps match the principles of social tourism? Social tourism has been especially defined as considering the holiday opportunities of low-income participants (Minnaert *et al.*, 2012, p. 20). This raises the main concern of this research and would argue for assuming another paradoxical rejection of camps from social tourism. For several decades, children's camps have become increasingly expensive; middle-class children in particular who cannot benefit from social funding are excluded from camps (Chauvin, 2002; Downs, 2009, p. 342), whereas low-income families can receive financial aids from the Family Allowance Fund, with the amount varying according to their incomes and the rules of each department (Brunel, 2013). More precisely, the different categories of social tourism (Minnaert *et al.*, 2011, pp. 406–407) are useful to better understand how children's camps can be connected, or not, to the criteria of social tourism. The French camps would correspond to the participation model or the adaptation model, with camps conceived for disadvantaged persons. Children in care can belong to these targets and models, and some of them experience camps thanks to social funding from public services. But the models cannot function completely, as the high-cost camps are also offered to well-to-do children. Thus, the current relations between offers and demands are quite different from the inclusion model identified by Minnaert *et al.*, because (i) the well-to-do class can search for separate costly camps; (ii) the middle class experiences financial difficulties in participating in camps; and (iii), as a result, low-income and disadvantaged children can experience the camps without social mixing, even though this can occur in some camps. This trend corresponds with the evolving meaning of the notion of solidarity (Blais, 2008).

This current situation questions the previous (inter)national models of social tourism. Social tourism has been implemented by unions and social tourism organizations (Diekmann and McCabe, 2011). However, Caire analysed the trajectory of French social tourism, from a 'cost-saving and non-monetary form of tourism' to the 'increasing industrialisation and professionalisation of social tourism' (Caire, 2012, pp. 76–85). We highlight that this trajectory corresponds very well to the changes in children's camps – as they are often organized by the same managers. Thus, the perspective already defined by Jean Froidure is confirmed. The two previous models that he stressed, Eastern Europe and Western Europe, disappeared and were covered by the Anglo-Saxon model, implementing free competition in a context of crisis and separated parts of society (Froidure, 1997, p. 178). However, several not-for-profit organizations were analysed, each choosing different positions in the socio-economical context. Nicourd (2009) highlighted that popular education organizations: (i) use a business model; (ii) implement public policies by benefiting from their financial support; or (iii) refuse the two previous approaches and assert their own identity thanks to volunteers. This leads us to the assumption that camp organizers can still be diverse according to these criteria, and that this diversity is no longer framed by national or supra-national borders. Is geographical discontinuity in today's children's camps a form of social tourism, or would it be framed according to another kind of distinction?

It appears that the current camps are distinguished from past camps by the great variety of themes (Chauvin, 2008, p. 146), corresponding to a marketed segmentation. All camps belonged to a business model. Several authors underline the phenomenon of de-differentiation between children's camps and family holidays, owing to the increase in family holidays with the children (OVLEJ, 2013). The educational specificity of the camp would be lost as the educational activities would extend to all forms of leisure time (Houssaye, 2002, p. 182). Leisure trends have developed in camps: walks and simple games have been replaced by 'costly activities such as sailing or horse riding' (Downs, 2009, p. 342). According to Gardet, these 'requirements' go hand in hand with less carefreeness and more individualism (Gardet, 2014, p. 13), which encourage expectations towards particular themed options. Greater concern for security is complemented by an expectation of intimacy (OVLEJ, 2005), which fosters improved comfort in the camps' accommodations. Generally, adolescents' expectations are divided between sport (OVLEJ, 2008) and cultural activities. In both cases, it is a question of actively engaging with the activities (Carr, 2006) and learning skills from practising them, which cannot be provided with only simple games. These criteria lead organizers to have increased costs and forced them to increase prices.

10.3 Characteristics of the Camps

10.3.1 Prices and (lack of) accessibility

The high prices of camps are considered as the first reason for disaffection with them (Buisson and Roure, 2013; Ménard, 2013). Beyond the development of themed activities, the increase in costs has been down to the improvement of the working conditions for camp leaders (Nutte, 2012). Ménard (2013) pointed out that partial financial support is necessary for middle-class families. Yet, it is estimated that only 65% of families are supported: 25% by works committee, 19% by a Family Allowance Fund, 14% by local authorities, and 7% by other funding

Table 10.1. Children's camps alongside different types of holiday. (Adapted from Ménard, 2013, p. 35.)

Type of holiday place	Average fee per day per child (€)
Stay abroad	115
Summer camp in France	63
Family holiday village	47
Low-cost summer camp in France or abroad	28–45
Day leisure centre	35
Bungalow in a campsite (based on a family with two parents and two children)	17–35
Five-star campsite (based on rented site for two persons)	20–25
Scout summer camp	15
One-star campsite (based on a rented site for two persons)	7–10

(Wittersheim, 2015). Thus, in this context, it would be this middle class that has to be situated as lower class according to the level of available money to pay for a camp. For their parents, how is it possible to pay a fee of an average of €63/day (Table 10.1), i.e. €440/week (Ménard, 2013), which often doesn't include transport costs, if they do not receive any financial support?

There is also lack of accessibility to faraway destinations and more expensive activities (Table 10.2). A number of camps with simple activities located in France are affordable at a price of around €60/week, but inequalities exist, with the middle class being deprived of the faraway camp locations, great tours abroad or even in France.

10.3.2 Locations, distances and relationship with the environment

Thus, social tourism has to be questioned not only in terms of prices and activities, but also considering the geographical relationships between the places of departure and arrivals and distances. If social tourism was characterized for a long time by a close relation between home and destination, the places mentioned in Table 10.2 show that this is no longer the case. However,

Table 10.2. Camp offers of the main French social tourism organizations and their prices. (From UCPA, 2016, 2018; VVL, 2018; PEP Alsace, 2018; PEP Découvertes, 2018; Ligue de l'enseignement, 2018; Colosolidaire, 2018)

Example of camps with a departure from Paris	Average fees per day per child (€)
Abroad	
Camp in Asia	160–200
Camp in North America	140–200
Camp in another European country(ies)	70–130
In France	
Mechanical sports camp	70–100
Musical, artistic, movie camp	60–100
Horse-riding camp	60–100
Water sports or mountain sports camp in France	50–90
Simpler camp with outdoor and team games, walks	50–60

a gradient is established according to different tourist belts (Miossec, 1977). For instance, in summer 2017 the Pupilles de l'Enseignement Public (PEP) camps, one of the main organizations running non-profit camps, were: (i) either in the countryside with locations in unknown villages but themed and attractive activities; (ii) located in mountain or coastal French resorts, which are more renowned for their social tourism identity with a simpler programme; or (iii) rolled out abroad also in popular destinations in the Mediterranean area (Michel and Moisy, 2017).

On the other hand, the relationship with the environment has to be stressed. Most of the camps, from the beginning, have been located in tourist areas with 'natural' landscapes (Michel, 2018, p. 104–105). However, we have to ask what value each organizer adds to the environment around the camp. This issue is connected to finance, because visual and material uses of the landscape in the surroundings permits costly activities to be avoided and educational objectives from nature to be defined (Dresner and Gill, 1994; Collado *et al.*, 2013, p. 43), when other camps focus only on the social space established by the gathering of participants (Alexandris and Kouthouris, 2005, p. 51; Dahl *et al.*, 2013, p. 109).

10.4 Methodology

The interviewees were chosen according to the camp prices offered: a threshold of €45/day/child was defined as a maximum, from the inventory of the range of offers by the French organizers. Thus, above €45, more camps are offered with more costly activities or even simpler activities, while below €45, the exploratory research permitted us to identify very few offers in France. We identified the organizations on the list of regional members of UNAT (the main French network of social tourism organizations). From the inventory of the camps available online, we included in the research almost all these regional members with at least one offer of a camp below €45/day/child, and which accepted our invitation to participate in this study. In the second stage, a snowball approach allowed us to find some other organizations outside the UNAT network. In particular, it was found that some sociocultural centres and youth and cultural centres also organize some low-cost camps. With this type of organization, the visibility of their offers is lower, as they do not have a 'commercial' approach, but they organize their camps more on a one-off basis as part of longer educational projects and mainly for their permanent youth members. Thus, while we can assume that we contacted most of the not-for-profit organizations that have a regular offer of camps on display in their catalogue and that present some low-cost offers, we also know that other not-for-profit organizations, among the sociocultural centres, also offer very economical camps. However, only two of them with visible offers online were interviewed. In all, 16 organizers were contacted, and agreed to be interviewed. They expressed their concern for the issue, which enables closer contact with them. Table 10.3 presents their characteristics. These organizers of low-cost camps are located in different regions in France. Most of them offer camps that are located in tourist areas: coasts, mountains, urban destinations, although some of them are located in areas in the countryside with low levels of tourism. Nine of them are

Table 10.3. Participants interviewed.

No.	French region	Type of summer camp location	Position within the organizational camp system	Other annual activities
1	Brittany	Mainly coastal	Organizer	Other stays and leisure activities
2	Brittany	Rural	Organizer and host	Farming
3	Normandy	Mainly coastal	Organizer and host	Other stays
4	Nouvelle-Aquitaine	Mainly coastal	Organizer	Leisure activities
5	Nouvelle-Aquitaine	Coastal	Organizer and host	Other stays
6	Nouvelle-Aquitaine	Coastal, rural and mountainous	Organizer	Other stays
7	Occitanie	Mountainous	Organizer and host	Other stays
8	Occitanie	Mountainous	Organizer and host	Few other stays
9	Occitanie	Coastal	Organizer and host	Other stays
10	Auvergne-Rhône-Alpes	Urban and rural	Organizer	No
11	Auvergne-Rhône-Alpes	Mountainous	Organizer	No
12	Bourgogne-Franche-Comté	Rural	Organizer and host	Other stays and leisure activities
13	Bourgogne-Franche-Comté	Mountainous	Organizer and host	Other stays and farming
14	Alsace	Rural	Organizer	Other stays and leisure activities
15	Alsace	Urban abroad	Organizer	Leisure activities
16	Centre-Val de Loire	Rural	Organizer and host	Other stays

both organizer and host, while seven of them are only organizers: in this case, they rent one or several accommodation units to operate their camps. Most of them have other annual activities to enable a permanent economy for their organization. These consist often of other educational stays, such as school residential trips, but also sometimes stays for disabled persons. The absence of other activities means either they do camps throughout the year, or they do camps just during the summer.

The interviews were conducted in October 2018. A common guide was defined and structured in three main parts:

1. The theme of motivations aimed at understanding the organizer's identity, and includes the historical background of the activities, in particular the low-cost camps, as well as the networks in which they participate.
2. The theme of the mode of production is focused on the way of organizing the low-cost camp(s), highlighting the approach of pricing, the parsimonious steps inside the organization and with the suppliers. Within this theme, we dealt in particular with the issue of volunteering to plan the camps.
3. The last theme looks into the objectives and benefits of the low-cost camps for children according to the interviewee, examining which objectives she/he emphasizes among the range of educational objectives: social life, coeducation, autonomy, freedom, human or more-than-human encounters and relationship with the environment. Every interview was conducted by phone and recorded. The duration of each one was between 20 and 50 min. All the contents were recorded and transcribed.

After transcription, the different items from the interview guide were used to classify the participants. However, as they do not constitute a 'pure sample' but are more original examples of organizing camps, we aimed to analyse their organizations using a comprehensive approach (Kaufmann, 2006). The analysis was carried out in two steps: the analysis of each interview and the themed analysis (Blanchet and Gotman, 2011). The first step to understand the individuals' positions was a focus on each interview, assuming that each individual holds a particular position. Within this first step, we identified the practices and the representations of each interviewee and we tried to link them. The second step consisted of going beyond the specificity of each interview. Thus, we searched for coherence identifying the similarities between different interviews. Classifying into a few ideal types was not possible, as the ways of conceiving camps are diverse and sometimes very specific. However, below we present a twofold structure in which the different organizers are placed.

10.5 Results

The inventory of the organizers of low-cost camps ends up bringing to the fore of the analysis a contrast between urban and rural organizers. Using the terms 'urban' and 'rural', we mean not only a difference in physical settings (organizers inhabiting and/or organizing camps in cities versus organizers inhabiting and/or organizing camps in the countryside), but a cultural relationship with urban or rural environments both in the practical and imaginary relationships between city and countryside (Debarbieux, 2005). In particular, we analysed what status they confer to the natural environment (Mathieu and de Lafond, 2016).

10.5.1 The low-cost camps of urban organizers: Social objectives for children and skilled management

Some organizers, either located in cities or in the countryside, conceive the objectives of camps as having an urban culture. They are focused on the benefits of a social life. Their camps are located mainly in cities or in coastal resorts. Some natural settings in the countryside may also be chosen, but just as locations to organize social activities or because there is cheap accommodation in the countryside. What matters more is gathering children together and an educational way of living in the camp by developing relationships with others and taking pleasure while engaging in the activities. To organize in this way, they benefit from management skills acquired from an urban sphere: higher education in business, urban social networks, knowledge of the urban stakeholders and their financial supports, and knowledge of young urban people. In their minds, this bundle of skills does not entail consideration of the natural environment to conceive the camps. Within this general method of organization, the differences come from the position of the organization.

The closer camps without a theme and the search for full accommodation

To take the example of some simple camps without a theme, among all the offers, lower prices can arise when associated with the ability to entirely fill the accommodation capacity. Thus lower camp prices were identified in an organization that chooses to fill its hosting capacities to the maximum: 'I prefer the accommodation to be full with low-cost camps and I can potentially compensate for my loss of profit margin as I have more people who come to our accommodation. I prefer filling one to 90% with camps costing 180 euros a week per person, rather than filling it to 30% with camps costing 500 euros a week per person' (interviewee no. 3). Beyond the management skills that enable this method, a human representation of camps has triggered this choice, stressing the need for simplicity: 'We removed paid activities, the camp leaders were put back at the core of the project and thus we reduced the cost of the camp. Basically, a camp includes transport, full board, and activities. Therefore, we have maintained some activities, the less expensive ones, and we have asked the camp leaders to find once again the core of their job' (interviewee no. 3).

It ensures that some camps are offered at €180/week. He points out: 'Children in care, thanks to the social support provided by the County Council, have the possibility of benefiting from two or three-week camps, with quite expensive prices, and costly activities. We observe a little of the inverse effect: the middle class, or even the well-to-do, choose the camp as a mode of custody' (interviewee no. 3). In this case, the low-cost camps cater for the middle-class demand, contributing to the model of inclusion; however, this requires solving the issue of differences of camp programmes in the same place and a social space that does not seem to include the search for otherness and natural environments.

International camps and the search for specific funding to reduce the selling price

Lower prices, at €30/day/child, are also found at a sociocultural centre located in a suburban area. This centre is very different from the previous organization, as it does not manage any accommodation. It benefits from public regular funding thanks to local authorities, and specific funding to support the camps themselves: 'I organized projects and responded to a call for proposals using the programme "*Fonds Publics et Territoires*" – I think they exist everywhere in France; they are supported by the Family Allowance Fund, to try to have an affordable price, for families and adolescents' (interviewee no. 15). Unlike the previous organizer, who implements private management skills commercially, this one is much more embedded in a public funding system. However, he must be skilled and proactive to successfully answer the calls for proposals and to obtain grants, and he also increases the support through self-funding actions: 'At the beginning of the year, we send mails to nearby shops, to districts, to see if there are odd jobs to be done by the adolescents. Thus they register for these jobs, there is a till that I manage here, and every time they have a self-funding action, they have money which accumulates in this till, and they choose how they spend it, for the camp to Paris, Berlin, Rome, etc. They really decide' (interviewee no. 15).

Among these resources that parents do not have to pay for, this organization includes themed activities: 'for the kids aged 10 to 13, if we have decided to organize a camp with water sports so obviously the cost will be more expensive. The activities cost more, thus the camp costs more. Or then I apply for grants, arguing that there will be water-skiing during a week' (interviewee no. 15). For adolescents, international camps are restricted to the main tourist urban destinations in Western Europe. Therefore, with similar costs in comparison with the previous organizer, a 'more-than-commercial' approach permits them to develop mobility, engage with different experiences and discover otherness.

Themed camps and the answer to adolescents' expectations in economical accommodation

A third subcategory of urban ways to organize camps can be found in three organizations that also do not manage their own accommodation. They express more the need to offer camps that correspond to the current societal expectations and therefore are themed. Below €45/day/child, they comprise a bracket of offers between €30 and 45. They search for moderate choices of costly themed activities: 'we choose less expensive activities, for instance we go sledging, but we don't go skiing, dog sledding or snowmobiling' (interviewee no. 14), or 'our prices are affordable because we don't have services delivery, we don't plan tree climbing, karting and sailing. We don't have those activities, a few at the outside, but it is rare' (interviewee no. 10). They are cautious not to carry out many activities; they seek rather to define a main theme to make the camp attractive: 'all the camps have a main theme, and we use outside service providers beyond camp leaders, because today we wouldn't be able to sell camps which would just be called "coastal summer camp". We are obliged to do it like that, thus our challenge is to keep prices affordable' (interviewee no. 1).

Their way of reducing the prices of these costlier choices is to lower accommodation costs: 'We know that the amount of 300 euros a week is reasonable. It means that we use often campsites and less expensive activities and we also own camping gear to have lower costs' (interviewee no. 1). Or: 'we have a financial agreement with the agricultural college where we organize the camps. The college is in favour of low-cost camps, because it is twice as cheap as the other college buildings we pay for during the summer' (interviewee no. 10). A third step to reduce the cost is operated by only one organization, which is able to charge only €30/day/child by using volunteer involvement to prepare and participate in the camp: 'we asked all our contacts to form a volunteer team, for cooking, service staff and camp leaders' (interviewee no. 14).

10.5.2 Rural ways of running low-cost camps: Social and environmental objectives for children and self-funding

In rural areas, the organizers are all hosts and often have other activities than camps during the year: school trips, or adult stays. They live in the rural camp setting or close by. They run camps as a very non-profit activity to allow children to have access to holidays. They do not mention the use of specific funding to support their low-cost camps. Their offers are very limited owing to low accommodation capacity as well as low demand for these rural camps. Yet, in the majority of cases they pay a great deal of attention to the objective of promoting a relationship with nature.

Farm camps in tents

Three farms were identified as organizing camps in tents. These farms either belong to the rural welcoming network '*Accueil paysan*' or they are independent. They aim to help children discover the natural environment and farm life and eat organic food. Some of their low-cost camps are offered at prices between €34 and 45/day/child. The children in the camps are accommodated in very simple stays with tents. This parsimonious way of running the camp is important, and, as the farm has other regular means of functioning, the summer camp is a one-time contribution during the year that does not aim to make a profit: 'we don't make money regarding our holiday activities so we can make them affordable for all. The organization takes no benefit from them and it's not the objective' (interviewee no. 2). They justify this option by arguing that they have other social tourism resources: 'we organize 300 horse-riding stays, that's how we earn, and on the basis of about ten forest camps, we don't need to make money' (interviewee no. 12). They also have agricultural resources thanks to the sale of farm products. It is also essential to notice the unpaid contributions of the farm employees to some tasks during the year.

The objective of developing a relationship with nature during the camps is strong but is often implemented with the possibility of choice and autonomy being given to the children, either according to the type of camp: 'not all children are interested in horse-riding; they are afraid, or don't know the animal. Therefore we imagined a less expensive camp where children can go into the forest to build tree houses, discover animals, and sleep in a tent' (interviewee no. 12), or during the camp: 'we encourage children to become more autonomous and more confident, to make a fire, to sleep under the stars, to choose their activities. The child can choose to make a water mill in the nearby river, or to brush donkeys' (interviewee no. 2). These principles are connected to seeking both human well-being and animal welfare in an environment that is new and different: 'our objective is really animal welfare, leading kids to discover animals, because we have many city dwellers, and they have never been in contact with a hen; they have never collected a real egg' (interviewee no. 12). While some organizers point out the contact with animals, another also insists on the benefit from the landscape: 'We are in a village located in a low mountain range. The mountain is just above us, where we can go hiking and our forest is two metres from the campsite' (interviewee no. 13). This mountain is the rural environment enabling camps to be planned in partnership with shepherds to locate the tents along a camp itinerary. This mobile implementation is also seen as economical: 'the adolescent camp in the mountain pasture is also cheaper, because we are autonomous at every meal' (interviewee no. 13). All these camps potentially share the objective of children's inclusion in rural environments: the camps are highly integrated into the local territory to enable discovery of their environment, which is a very different position from the 'urban ways' of camps explained above.

The farm camps often present the highest degree of relationship to nature for children and are associated with the discovery of a cultivated milieu. The other types of rural camps described below aim at a lesser degree of this relationship; however, they do not ignore this issue, which contributes in other ways to reducing the costs of camps.

Repeated nature camps for loyal children and adolescents

Three organizations without permanent salaried staff were identified with an offer of camps at prices between €37 and 42/day/child. Two of them manage accommodation, or rent it, both in a natural environment. To enable sustainability, they highlight two favourable factors. First, the absence of salaried staff entails contributions from volunteers: 'we are about 20 volunteers, which reduces the costs' (interviewee no. 8). Owing to the lack of other activities over the year, they insist on parsimonious and balanced camps: 'our camps are self-financing, and the unpaid organization allows that easily' (interviewee no. 11). Second, they can benefit from loyal users, either coming from far, or nearby: 'for some years, the 'Secours Populaire' team has been interested in our camp, which we use to welcome kids who come from the Paris region'. Otherwise, it is more individualized: 'we have kids identified in the schools where we are working, or identified by the school social worker in another school, often through direct contacts. Most of our registrations are done like that, through personal networking' (interviewee no. 11). If this loyalty is not only a rural feature, it is favoured by the framework of the rural territory, and can bring about repetition and renewal of the children's contributions: 'our educational project aims to welcome children from nearby families and to gain their loyalty. We ensure that they are satisfied and happy, and that they come back the following years, up to the age of 15 or 16 if they wish. Thus we try to gain their loyalty by this means, and we try to lead them to train to become camp leaders' (interviewee no. 8).

The first nature camp for children

Apart from farms, another type of low-cost camp is organized by host rural or coastal centres in a natural environment, and they aim at welcoming children to the camp for their first time outside of their family. Their prices are between €39 and 44/day/child. Unlike the camp discussed above their objective is not repetition of the stay at the same location. The first camp for a child is more conceived as an initial step before experiencing other camps in other settings. This first camp is certainly at a higher cost and/or is supported by social services. The social dimension of this type of camp is strong: 'our idea is to ensure that kids who have never gone out of the working-class area of their town, their public housing, their troubled area, can go on holiday, but above all they can have contact with nature' (interviewee no. 7). More than the offers by the farms or other rural centres, these focus on the social-natural objective. They welcome children in care and middle-class children together: 'I think there is a financial issue for parents. We have more and more children who come for only one week, and more and more children who come from the middle class. They choose the low-cost camp because the parents want them to be left alone, but financially they make them leave for only one week' (interviewee no. 5). This diversity of publics leads to a demand for seeking contact with nature for the different social classes.

Beyond the financial issue, the first experience of the great outdoors is emphasized: 'the idea consists of organizing camping, because it calms the kids' (interviewee no. 7, explaining a process related to the therapeutic landscape) (Gesler, 1992). Awareness of this process has been triggered by training from which the current organizer benefits delivered by an instructor dealing with environmental education. Thus, beyond the framework of a social group, opening up to the natural world is stressed as something that is not cheap to implement when and where it is experienced in the great outdoors: 'Nature activities enable children to discover, to have emotions and to live things. Going underground is a very strong activity, or climbing a tree and sleeping in a hammock with roe deer passing below, and listening to owls. It's a total immersion in nature which is not at all at a lower cost' (interviewee no. 7). The ability to offer this camp price comes from other financial resources during the year: other children's camps, school residential trips, and so on.

Regional rural travelling camps

The last type of rural organizers are those who are not hosts, but who send children and adolescents especially outside of the rural milieu. The organizer of a sociocultural centre that receives some support from public authorities and an independent organization were interviewed.

Their lower prices are in the upper bracket of our research, between €43 and 45/day/child. They seem to face the expectation of some 'consumer activities' for (pre)adolescents: 'there are consumer activities to attract and include what adolescents want to do [...], our adolescents wish to go to the coast. At some point we organize the activities around this, and there are these theme parks as well' (interviewee no. 4). The themed attraction is confirmed by the other organizer: 'we organize everything by ourselves; we don't use external service contractors, except for the flagship activity of the camp, because we have to provide a flagship activity which goes down very well' (interviewee no. 6).

As they are not hosts, the common option to reduce the cost is the use of cheap accommodation: 'we try to find hosts who can offer correct and affordable prices, which means campsites, natural sites, and hosts who accept self-management and who can offer the accommodation at 20 euros per night per person' (interviewee no. 6), in particular when it is only summer camps: 'we manage to maintain affordable prices during the summer, because the camp is in a tent' (interviewee no. 4). The other option is a very original negotiation process with the parents: 'we are interested in the middle class who do not go on holiday. We organize our own grants to help families, to offer the most affordable price for this population' (interviewee no. 6).

The levers to decrease the high prices of children's camps depend on the level of financial support, which is linked to the geographical and social milieu in which the organizer is encompassed: some 'urban' organizers can be prompted to apply for and benefit from particular grants as well as to gather a great number of volunteers for many camps. In the rural milieu, the organizers can take advantage of their host status – if they have this – to welcome diverse users, and to highlight the natural value of their setting (Table 10.4). In the case of weak public support directly to the camp, the factors in achieving a lower price are the short distance between home and camp, the mode of accommodation – in particular camping – the choice of simple activities and volunteering.

10.5 Conclusions

Every organizer interviewed within this research develops some of the factors leading to low-cost camps. However no one looks at adding them all together, because everyone wants to offer some amenities: an attractive activity or comfortable accommodation or organic food. Thus, in principle, the camp prices could be even lower than these low costs, yet they would probably lose their values, objectives and attraction. However, every organizer develops a way of conceiving camps, which is evidence that reducing the cost of camps is feasible. They contradict the majority of organizers, who think that it is impossible to organize children's camps with low-cost options. Several implications can be gained from this analysis.

First, the pressure to offer themed and expensive activities has to be questioned. The example of the flagship and expensive activities is revealing: to what extent are they necessary to the attraction of the camp? Organizers feel forced to plan them, but there are no results in this context providing evidence of their necessity.

Second, in the case of a limited camp activity by small not-for-profit organizations, volunteering is useful, if not necessary, to prepare and implement camps. In this way, earlier organizations of camps with little paid workforce are renewed. Without this volunteering, some rural organizers would no longer exist.

Third, the camps have to be integrated into the environment: this integration enables benefits thanks to the economic exchanges with other stakeholders in the territory. These local networks favour cheaper economic models for camps.

Fourth, regarding the experience of children, it is not necessary to go far to make great discoveries. The value of the camp has to be assessed more in relation to the child's worldview, referring to the need to develop childhood studies in this context of children's camps. Eventually, beyond the socioeconomic criteria of social tourism, slow and educational forms of tourism have to be more included in the organization of children's camps.

Table 10.4. Results: the systems of low-cost camps.

Main frame of camps	Social space			Social and environmental context			
Ways of reducing the cost	Camp without theme	International discovery camp	Themed camp	Farm camp	Repeated nature camp	First nature camp	Regional travelling camp
Prices (€) for low-cost camps	28	30	30–45	34–45	37	39–44	43–45
Volunteering to organize the camps	Weak	Very strong	Weak	Strong	Very strong	Strong	Strong
Public funding towards the camps	Weak	Strong	Weak	Weak	Weak	Weak	Weak
Costly themed activities	No	Yes	Yes	No	No	Yes or No	Yes
Search for contact with nature	Weak	Weak	Weak	Very strong	Strong	Strong	Moderate
Search for otherness	Weak	Strong	Weak	Moderate	Moderate	Moderate	Moderate

Acknowledgements

We are grateful to the organizers of the camps for their attention to the research and their agreement to the use of the interviews for this publication. We thank in particular Kevin Magner for his proofreading.

References

Alexandris, K. and Kouthouris, C. (2005) Personal incentives for participation in summer children's camps: Investigating their relationships with satisfaction and loyalty. *Managing Leisure* 10, 39–53.

Blais, M.-C. (2008) La solidarité. [Solidarity.]. *Le Télémaque* 33, 9–24.

Blanchet, A. and Gotman, A. (2011) *L'enquête et ses méthodes L'entretien. [The Survey and its Methods: The Interview.]*. Armand Colin, Paris.

Bocquet, J.-M. and Dheilly, C. (2018) Les colonies de vacances seraient-elles vraiment devenues du tourisme?! [Would summer camps really have become tourism?!]. Available at: http://theconversation.com/les-colonies-de-vacances-seraient-elles-vraiment-devenues-du-tourisme-95167 (accessed 29 November 2019).

Brunel, J. (2013) Vacaf, la montée en puissance d'un service public. [Vacaf, the rise of a public service.]. In: *Des colos innovantes!, Actes du colloque enfants & ados*. Unat, Paris, pp. 79–81.

Buisson, C. and Roure, E. (2013) Lutter contre la fracture touristique. [Fight against the tourist divide.]. Contrôle général économique et financier. Paris. Available at: https://www.economie.gouv.fr/files/rapport_lutter_contre_fracture_sociale_buisson_tome1_0.pdf (accessed 19 October 2020).

Caire, G. (2006) Les associations françaises de tourisme social face aux politiques européennes. [French social tourism associations facing European policies.]. *Revue internationale de l'économie sociale* 300, 30–44.

Caire, G. (2012) Social tourism and the social economy. In: McCabe, S., Minnaert, L. and Diekmann, A. (eds) *Social Tourism in Europe Theory and Practice*. Channel View Publications, Bristol, UK, pp. 73–88.

Carr, N. (2006) A comparison of adolescents' and parents' holiday motivations and desires. *Tourism and Hospitality Research* 6(2), 129–142.

Chaudière, C. and Baldacchino, J. (2018) Les jolies colonies de vacances sont-elles en danger? [Are pretty summer camps in danger?]. Available at: https://www.franceinter.fr/societe/les-jolies-colonies-de-vacances-sont-elles-en-danger (accessed 29 November 2019).

Chauvin, J. (2002) *Le tourisme social et associatif en France. [Social and Associative Tourism in France.]*. L'Harmattan, Paris.

Chauvin, J. (2008) *Les colonies de vacances: Domaine privilégié de l'éducation populaire. [Holiday Camps: Privileged Area of Popular Education.]*. L'Harmattan, Paris.

Choudhary, V. (2010) *Educational Tourism*. Centrum Press, New Delhi.

Collado, S., Staats, H. and Corraliza, J.A. (2013) Experiencing nature in children's summer camps: Affective, cognitive and behavioural consequences. *Journal of Environmental Psychology* 33, 37–44.

Colosolidaire (2018) *Colosolidaire: Le réseau d'acteurs solidaires qui favorise le départ en colonies de vacances pour tous les enfants. Catalogue Eté 2018. [Colosolidaire: The network of solidarity actors that promotes the departure to summer camps for all children. Summer Catalogue 2018]*. Association Colosolidaire, Bordeaux, France.

Dahl, T.I., Sethre-Hofstad, L. and Salomon, G. (2013) Intentionally designed thinking and experience spaces: What we learned at summer camp. *Learning Environment Resources* 16, 91–112.

Debarbieux, B. (2005) Obsolescence ou actualité des objets géographiques modernes? A propos de la ville, de la campagne, du périurbain et de quelques autres objets géographiques conventionnels. [Obsolescence or topicality of modern geographical objects? About the city, the countryside, the peri-urban area and some other conventional geographical objects.]. In: Arlaud, S., Jean, Y. and Royoux, D. (eds) *Rural urbain Nouveaux liens, nouvelles frontiers*. Presses Universitaires de Rennes, Rennes, France, pp. 33–43.

Diekmann, A. and McCabe, S. (2011) Systems of social tourism in the European Union: A critical review. *Current Issues in Tourism* 14(5), 417–430.

Downs, L.L. (2009) *Histoire des colonies de vacances de 1880 nos jours. [History of Summer Camps from 1880 to the Present Day]*. Perrin, Paris.

Dresner, M. and Gill, M. (1994) Environmental education at summer nature camp. *Journal of Environmental Education* 25(3), 35–41.

Froidure, J. (1997) *Du tourisme social au tourisme associatif: Crises et mutations des associations françaises de tourisme. [From Social Tourism to Associative Tourism: Crises and Changes in French Tourism Associations.]*. L'Harmattan, Paris.

Fullagar, S., Markwell, K. and Wilson, E. (eds) (2012) *Slow Tourism. Experiences and Mobilities*. Channel View Publications, Bristol, UK.

Gardet, M. (2014) *Les colonies de vacances. [Summer Camps.]*. Le cherche midi, Paris.

Gesler, W. (1992) Therapeutic landscapes: medical issues in the light of the cultural geography. *Social Science & Medicine* 34(7), 735–746. DOI: 10.1016/0277-9536(92)90360-3.

Houssaye, J. (1998) Le centre de vacances et de loisirs prisonnier de la forme scolaire. [The vacation and leisure center trapped in the school form]. *Revue française de pédagogie* 125, 95–107.

Houssaye, J. (2002) Les centres de vacances: la fin des finalités. [Holiday centres: The end of the ends]. *Education et francophonie* 30, 172–186.

Kaufmann, J.-C. (2006) *L'entretien compréhensif. [The Comprehensive Interview]*. Armand Colin, Paris.

Ligue de l'enseignement (2018) *Colonies Vacances pour tous Eté 2018 4–6 ans, 6–11 ans, 12–14 ans. [Summer camps holidays for all summer 2018 4–6 years, 6–11 years, 12–14 years]*. Ligue de l'enseignement, Paris.

Mathieu, N. and de Lafond, V. (2016) Sustainable futures and rural-urban relationships in rural areas under urban pressure. In: Mathieu, N. (ed.) *Les relations ville-campagne. Histoire d'une question politique et scientifique*. L'Harmattan, Paris, pp. 177–196.

Ménard, M. (2013) Rapport d'information sur l'accessibilité des jeunes aux séjours collectifs et de loisirs. [Information report on the accessibility of young people to group and leisure stays]. Assemblée nationale, Paris. Available at: www.assemblee-nationale.fr/14/rap-info/i1236.asp (accessed 29 November 2019).

Michel, X. (2018) Les centres de séjour du tourisme social. [Social tourism accommodation centres]. In: Guillemin, P. (ed.) *Atlas de la Manche Des polders au pôle d'air. Orep Editions*. Bayeux, France, pp. 104–105.

Michel, X. and Moisy, L. (2017) Spatialités et modes d'habiter du tourisme social. [Spatialities and ways of inhabiting in social tourism.]. In: Coëffé, V. (ed.) *Le tourisme. De nouvelles manières d'habiter le Monde*. Ellipses, Paris, pp. 178–196.

Minnaert, L., Maitland, R. and Miller, G. (2011) What is social tourism? *Current Issues in Tourism* 14(5), 403–415.

Minnaert, L., Diekmann, A. and McCabe, S. (2012) Defining social tourism and its historical context. In: McCabe, S., Minnaert, L. and Diekmann, A. (eds) *Social Tourism in Europe: Theory and Practice*. Channel View Publications, Clevedon, UK, pp. 18–30.

Miossec, J.-M. (1977) Un modèle de l'espace touristique. [A model of tourist space]. *L'Espace géographique* 6(1), 41–48.

Nicourd, S. (2009) Education populaire: Organisation du travail associatif et action publique. [Popular education: Organization of associative work and public action.]. *Entreprises et histoire* 56, 62–72.

Nutte, A. (2012) Rapport du groupe de travail sur le contrat d'engagement éducatif. [Report of the working group on the educational employment contract.]. Inspection générale des affaires sociales. Paris. Available at: http://jeunes.gouv.fr/IMG/pdf/CEE_rapport_final_GT-2.pdf (accessed 29 November 2019).

Obrador, P. (2012) The place of the family in tourism research: Domesticity and thick sociality in the pool. *Annals of Tourism Research* 39(1), 401–420.

OVLEJ (2005) *Centres de vacances, colonies de vacances… Perceptions et attentes des parents. [Vacation centers, summer camps… Parents' perceptions and expectations]*. 10–15. Bulletin de l'OVLEJ, p. 21. Available at: www.ovlej.fr/wp-content/uploads/Bulletin_Ovlej_10_15.pdf

OVLEJ (2008) *Entre soi ou avec les autres. Séjours familiaux sans les parents ou colos. [Between oneself or with others. Family stays without parents or vacation camps… Parents' perceptions and expectations]*. 19. Bulletin de l'OVLEJ, p. 19. Available at: www.ovlej.fr/wp-content/uploads/Bulletin_Ovlej_19.pdf

OVLEJ (2013) *Le départ en vacances des enfants et des adolescents aujourd'hui: Progression des inégalités et resserrement autour de la famille. [Children and adolescents going on vacation today: Growing inequalities and tightening around the family]*. 41. Bulletin de l'OVLEJ, p. 12. Available at: www.ovlej.fr/wp-content/uploads/Bulletin_Ovlej_41.pdf

PEP Alsace (2018) *Les PEP Alsace: La big aventure 4–11 ans/12–17 ans Eté 2018. [PEP Alsace: The big adventure 4–11 years old/12–17 years old Summer 2018.]*. PEP Alsace, Colmar/Strasbourg, France.

PEP Découvertes (2018) *Catalogue Eté 2018: Mille et une façons de grandir! Colos et mini-séjours pour les 4–17 ans. [Catalog Summer 2018: A thousand and one ways to grow! Colos and mini-stays for 4–17 year olds.]*. PEP Découvertes, Créteil, France.

Ritchie, B.W. (2003) *Managing Educational Tourism*. Channel View Publications, Clevedon, UK.

Small, J. (2008) The absence of childhood in tourism studies. *Annals of Tourism Research* 35(3), 772–789.

UCPA (2016) *UCPA Odyssée Printemps/Eté/Automne séjours sportifs 2016. [UCPA Odyssey Spring/Summer/Autumn sports trips 2016]*. UCPA, Paris.

UCPA (2018) *UCPA Odyssée Séjours sportifs Hiver/Printemps 2019. [UCPA Odyssey Winter/Spring 2019 sports trips.]*. UCPA, Paris.

VVL (2018) *Séjours enfants juniors 2018 jeunes/linguistiques/familles. [Youth/language/family junior stays 2018]*. VVL, Vitry-sur-Seine, France.

Wittersheim, A. (2015) Panorama et baromètre des colonies de vacances. [Panarama and barometer of summer camps.]. In: *Les colos, un enjeu pour la jeunesse, Actes du 4ème colloque vacances enfants-ados*. Unat, Paris, pp. 24–29. Available at: https://www.unat.asso.fr/nos-actions/produire/actes-du-colloque-les-colos-un-enjeu-pour-la-jeunesse

11 Social Tourism for Low-Income Families: Challenges and Practical Implications in Pursuing Social Innovation

Joana Lima[1]* and Celeste Eusébio[2]

[1]Social Sciences School, University of Évora and CIDEHUS Research Unit, Évora, Portugal; [2]Department of Economics, Management, Industrial Engineering and Tourism, University of Aveiro and GOVCOPP Research Unit, Aveiro, Portugal

11.1 Introduction

Households in developed countries, and particularly in the Portuguese context, suffered significant impacts in the global economic recession after 2008 and are once again experiencing similar impacts as a result of the Covid-19 pandemic. Among these impacts can be highlighted a decrease in demand for work resulting in reduced working hours and increased unemployment, decreased ability to manage income in the face of rising living costs and debts incurred, and increased uncertainty and fear about the future. These circumstances place greater emphasis on the need to develop social policies that provide mechanisms for social innovation, such as the access of families to tourism.

In this context, the recognition of the benefits that tourism may have and the growth of tourism in recent decades do not prevent the existence of several constraints to tourism. Low income is considered the greatest of those constraints, making tourism inaccessible for many groups in society, even in developed economies. Thus, social tourism appears as an attempt to restore equal access to tourism.

Including social tourism programmes targeted at low-income families in social policy arises as one possible measure to combat the inequality of access to tourism. However, this measure requires research on the actual level of participation in social tourism programmes, on the interest that the low-income families may have in these initiatives and on the guidelines these programmes should respect in order to both be attractive for this segment and maximize the benefits for the families and society. Notwithstanding, few studies analysing tourism for low-income groups have been undertaken, this gap being more obvious if social tourism for low-income families is considered.

The present chapter aims to reduce the identified gap, discussing the importance of social tourism programmes for families and giving an analysis of the Portuguese situation. Concretely, this chapter presents an analysis of the visions of a sample of Portuguese families regarding this kind of offer and ends with a discussion regarding guidelines that social tourism programmes for families should follow in order to combine the maximization of its benefits and its attractiveness as a tourism product.

*Corresponding author: jisl@uevora.pt

 Social Tourism: Global Challenges and Approaches (eds J. Lima and C. Eusébio)
DOI: 10.1079/9781789241211.0012

11.2 The Potential of Social Tourism to Reduce Social Exclusion

Present-day society has suffered significant effects from the global economic recession after 2008 and the current Covid-19 pandemic situation. Households, even in developed countries, and particularly in the Portuguese context, have experienced consequences that must also be borne in mind when reflecting on social tourism. Sumner *et al.* (2020) estimate that Covid-19 will have a significant impact on global poverty that could increase for the first time since 1990 and that such an increase may represent a setback of approximately a decade in the world's progress in reducing poverty. The intensity of these effects will depend on how long the pandemic lasts, and on how national governments and the international community react (Sumner *et al.*, 2020). However, unemployment is one effect that will certainly impact the families around the world (and indeed is already visible) (ILO, 2020).

Unemployment can have a negative effect on people's mental health, resulting not only from the negative impacts of unemployment on family income, but also from the impacts in terms of social isolation (Clark, 2003; Paul and Moser, 2009). In addition, the psychological impacts of unemployment on parents can have a negative impact on parenting and consequently on children's well-being (Gray *et al.*, 2009). The experience of previous recessions indicates that involuntary unemployment typically has a negative impact on the physical and psychological health of the individuals affected by it, and other family members, including children, may also be affected. There are studies that analyse the effects of unemployment on children of families with different income levels, observing that children living in low-income families are more affected by the loss of a parent's job than children from families with higher income levels (e.g. Coelli, 2005 cited by Gray *et al.*, 2009).

In the past decade there has also been more widespread recognition of the harmful effects of social inequalities, reflected, for example, in the report by Standard & Poor's (2014), a market-oriented institution, which warned of the negative impact that social inequalities have in the process of economic growth. The Sustainable Development Goals, stablished by the United Nations (UN) in 2016, start with 'Goal 1: No Poverty' by 2030 and orient countries to 'mobilise efforts to end all forms of poverty, fight inequalities and tackle climate change' (UN, 2020). In the European Union (EU), concerns about the eradication of poverty are also evident, and in the year 2000, European leaders established a framework for strategic development and political coordination between Member States on issues related to poverty and social exclusion, in which Member States recognized non-governmental organizations (NGOs) (e.g. the European Anti-Poverty Network – REAPN), social partners and local and regional authorities as key players in this process (EC, 2010).

The priority of fighting poverty and social exclusion mentioned above was reiterated by REAPN (2008), underlining the need to rethink social protection, ensuring a fairer distribution of wealth. In this study, REAPN, among others, draws the following conclusions:

1. The difficulties and constraints to the social inclusion of people with lower incomes arise with the difficulties in accessing information.
2. Public services, among others, are not, in themselves, effective mechanisms for promoting social participation.
3. 'it is necessary that people acquire or recover levels of integration in social networks, provided by the family, work and/or other structures related to the different dimensions of social life', since these networks/structures are those that ensure conditions and enhance higher levels of participation/inclusion (REAPN, 2008, p. 7).

All these objectives can find tourism as an instrument to support their achievement. As mentioned in the Introduction to this book, since the 1990s, studies have concluded that access to holidays outside the usual environments contributes to personal and social development, family cohesion and the well-being of individuals (ETB, 1976; Smith and Hughes, 1999; EESC, 2006; Lehto *et al.*, 2009; Minnaert *et al.*, 2009, 2011; Alexander *et al.*, 2010; McCabe *et al.*, 2010; Lima *et al.*, 2018). This group of authors considers that tourism should be accessible to all who want to practise it and it should fall to the agents of supply (in which the public sector is also included) to guarantee this equal access to tourism. Following this line of thought, EESC (2006) states that everyone, including the less privileged elements of society, has the right to leisure time that allows the development of their personality and social integration, and that tourism is the practical expression of that right.

However, there are constraints that make the practice of tourism inaccessible to a large part of the population, with low income being one of the most important. The identified constraints engender high penalties, since the benefits of tourism are recognized both by the agencies that promote it and by the public authorities that integrate social tourism initiatives into their policies (Minnaert *et al.*, 2009).

Tourism is recognized as an activity that promotes access to information, and is an activity that, when planned, can foster a strong connection/cooperation between the public and private agents of a society, promoting efficiency in that same society. It is an activity that promotes socialization, through interaction between communities and visitors, as well as between the different elements that make up the travel group, thus also promoting learning processes (Lima, 2015).

In view of the serious issues that poverty and social exclusion are known to cause for societies and in the search for solutions to the problem, some initiatives have been put into practice both in the field of theoretical research and social intervention. As a result of the concern with looking at poverty and social exclusion as comprehensive and multidimensional concepts, the EU developed a measure that includes indicators related to income and non-monetary indicators of living standards, allowing a broader picture to be drawn of poverty and social exclusion – the material deprivation rate. The material deprivation rate defines material deprivation as the forced absence of at least three of nine items,[1] among which is the ability 'to go on holiday' (Wolff, 2009; Eurostat, 2018). The inclusion of this item substantiates the recognition of tourism as a need associated with the current lifestyle and an important dimension of today's society.

In fact, according to Smith and Hughes (1999), tourism should be considered a need associated with the current lifestyle and a dimension of the current society. In this context, not having access to holidays outside the place of residence and to leisure activities can be seen as a form of exclusion from the common lifestyle and from the needs that any individual must be able to satisfy in order to realize his or her right to be a complete citizen (Smith and Hughes, 1999). Thus, the inability to practise tourism, as an inability to share the lifestyle associated with today's society, may be a dimension of poverty (Smith and Hughes, 1999). These authors refer to the idea that poverty limits the ability of individuals to be active and complete members of society and conclude that the recognition of this relationship by some governments was the basis for the creation of social tourism programmes integrated in their social policies, such as France, Spain, Portugal or the UK, among others.

Another result of the concerns about social exclusion, social inequalities and the creation of mechanisms to combat or minimize these situations in partnership with social and local agents was the recognition of the importance of the social economy in the 1980s, with the promotion of a set of European Social Economy Conferences by the EU. The social economy finds its foundations in the search for a balance between economic growth and social cohesion, counting, for this purpose, on the actions developed by cooperatives, associations of social solidarity and mutualists (Namorado, 2004).

Considering the challenge of current social exclusion, social policies to combat behaviours that promote exclusion are more effective than redistributive or job access policies (Levitas, 2004; Sousa *et al.*, 2007; Minnaert *et al.*, 2009). Tourism represents an opportunity to change behaviours of exclusion, through participation in a dimension of present-day society that implies social interaction and contact with new contexts and realities. Participation in tourism activities promotes cultural exchange, relaxation, personal and social development, and behavioural changes. If tourism programmes are planned and executed to promote behavioural changes that perpetuate the cycle of poverty and exclusion of disadvantaged classes, the objective of social policies to combat exclusionary behaviours present in the most disadvantaged social classes can be achieved more effectively (Minnaert *et al.*, 2009). In social support for these families, guidance for activating their skills will be essential (Sousa *et al.*, 2007). The worsening of poverty and social exclusion is, therefore, one more factor that stresses the need to develop social policies that provide mechanisms of social innovation, as is the case of facilitating the access of families to the practice of tourism.

For the implementation of measures that really promote the insertion/participation of people in society, it is considered that social policies which are oriented to changing behaviours that promote exclusion, defined as 'new generation' policies (Sousa *et al.*, 2007), may integrate social tourism programmes as an instrument to combat exclusion (Minnaert *et al.*, 2009, 2011; McCabe *et al.*, 2010). The merit of these programmes is their ability to increase social innovation, for example, boosting learning processes through the experiences lived by low-income families, contributing to combating behaviours that perpetuate social exclusion (Minnaert *et al.*, 2009, 2011; McCabe *et al.*, 2010).

Durkheim mentioned that the intensity of the collective group depends on the density of the family group. This thought and the fact that a consequence of participation in social tourism programmes for families is the strengthening of family cohesion and proactivity (Minnaert *et al.*, 2009; McCabe *et al.*, 2010; Lima *et al.*, 2012; Lima, 2015), lead to the conclusion that it will fall to the agents involved in the definition of social policies to develop mechanisms that make it possible to take advantage of this change of environment, maximizing the return of those policies for society (Minnaert *et al.*, 2009; McCabe *et al.*, 2010).

Integrating tourism programmes into social policies is a measure that, despite arousing the interest of several entities, has not received effective recognition or research. In Portugal, there have been several social tourism programmes (financed by the state), targeted at the four segments (seniors, young people, low-income families and disabled people). However, nowadays the offer is almost non-existent.

11.3 Methodology

Considering the aim of the research and the few previous existent studies, a mixed methodology was chosen. First, a survey of Portuguese families ($N = 1165$) was undertaken, in order to analyse the level of participation and the interest of different types of families in social tourism programmes. Second, in-depth semi-structured interviews were conducted with 15 Portuguese low-income families (at least one parent and one child of each family), in order to understand their opinions regarding social tourism programmes for families. The option for a mixed methodology is particularly suitable for studies of an exploratory nature, with the qualitative approach seeking to answer the questions that emerged from the literature review carried out and complementing any quantitative studies.

11.3.1 Survey of Portuguese families

A survey was carried out with a sample of Portuguese families ($N = 1165$) to obtain information regarding the participation and level of interest of families in social tourism programmes. The population defined for this study was Portuguese families with at least one child aged between 6 and 17 years old and with different socioeconomic characteristics, resident in the central region of Portugal. The study population includes two types of families: families who have taken family holidays away from home in the past 4 years (FH families) and families who have not taken family holidays away from home in the past 4 years (NFH families).

Since it was not possible to survey all the families residing in the studied region, a multistep cluster-sampling approach was undertaken. The choice of this sampling technique was based not only on the objectives of the study, but also on the characteristics of the population under study (large and dispersed, in spatial terms). The sample included households from four municipalities – two inland and two in coastal areas, regions with higher levels of economic development (coastal areas), and regions facing demographic and economic problems (inland regions).

In the first phase, two NUTS III areas of the central region were selected, within which two municipalities were randomly selected. A group of schools was randomly chosen within each municipality and within each group of schools, two classes of each year of education (from 1st to 12th grade) were randomly selected. In each selected class, questionnaires were distributed to all students so they could ask their parents or legal guardians to fill it in. The distribution of questionnaires to the schools took about 2 months (between April and June 2014) and 2077 questionnaires were distributed to be filled in by the parents or legal guardians of the students. Some 1165 questionnaires (representing 825 FH families and 340 NFH families) were returned completed and were considered, which represents a response rate of around 58.5%.

Inside each group of families (FH and NFH), each family was classified according to its economic status. For this, the index of material deprivation was used. 'Deprived families' (or 'families with deprivation') are those who suffered forced absence of at least three of nine items that comprise the material deprivation rate and 'other families' (or 'families without deprivation') are those that do not suffer material deprivation.

11.3.2 Interviews with a sample of low-income families

Semi-structured interviews were conducted with a sample of low-income families residing in the municipality of Covilhã in order to understand their opinion regarding social tourism programmes for families. In the approach to low-income families, it was decided to restrict the geographic area of the study to residents in the municipality of Covilhã, despite the efforts and the time invested to be able to interview more families, also resident in other municipalities. As in the survey, the sample includes two types of families: families who have taken family holidays away from home in the past 4 years (FH families) and families who have not (NFH families).

At this stage of the study and considering the need for a definition that would allow the selection of families to be interviewed in the schools, a low-income family was defined as a family with subsidized children in schools (i.e. in the school year 2013/2014, families with reference annual income up to €5869.08).

Intentional non-probabilistic sampling was the chosen sampling method, as it is one of the most used in qualitative studies (Yin, 2011) and it allows the selection of cases to be analysed with the aim of obtaining a diversity of points of view (Patton, 1990; Yin, 2011). However, the concern to avoid bias, both in the design of the sample to be used and in the selection of the cases to be studied, was always present. Thus, of the various intentional sampling techniques that exist, we opted for the 'snowball'-type technique to approach low-income families. The interviews were carried out with one of the parents and at least one of the children, aged between 6 and 17 years old. All interviews were transcribed and subject to content analysis. A comparative analysis was carried out, trying to identify consistencies and inconsistencies between different observations (McCracken, 1988).

11.4 Results

11.4.1 Families who have taken family holidays away from home in the past 4 years

When asked about their living conditions, particularly in relation to material deprivation, the responses of the FH families surveyed permit the conclusion that 18.7% of these families were materially deprived, meaning that 18.7% of surveyed households suffered forced absence of at least three of nine items that comprise the material deprivation rate. In this study, as mentioned in the previous section, these households will be considered the group of 'deprived families' (N = 154) and the families that do not suffer material deprivation will be designated as 'other families' (N = 669).

Almost all of the respondents (97.6%) never participated in social tourism programmes, noting that 0.6% of the respondents did not answer this question and some of the 1.8% who claimed to have participated in social tourism programmes do not seem to have understood the concept well, despite being presented in the questionnaire. This hypothesis is raised since, when asked about the promoter of the programme in which they participated, they mention institutions that do not usually have the activity of promoting these types of programmes, such as, for example, private travel agencies and exclusive holiday clubs. This result is also, as mentioned by Lima (2015), a consequence of the absence

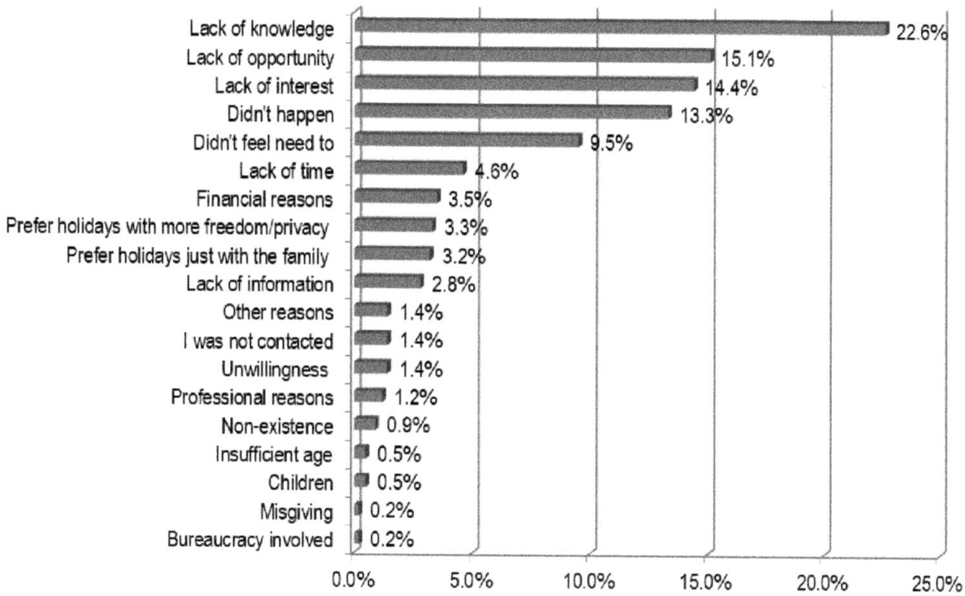

Fig. 11.1. FH families' motives for not participating in social tourism programmes.

of social tourism programmes for families with children in Portugal.

When questioned about why they have never participated in these types of programmes, respondents mostly refer to (22.6%) the lack of knowledge about their existence as the main reason, with 0.9% of respondents even mentioning the non-existence of these programmes (Fig. 11.1). A total of 14.4% of respondents reveal a lack of interest in this type of product as the reason for not participating – a reason that may be associated with preferring to go on holidays with more freedom/privacy (3.3%) or just with the family (3.2%). It is noteworthy, however, that 3.5% of respondents mention financial reasons as a reason for not participating, even considering the type of trips that social tourism programmes represent. Finally, the reason pointed out by 1.4% of respondents was 'I was not contacted', a reason that denotes some passivity and assumption that this type of programme should be proposed to potential participants.

Finally, regarding their interest in participating in social tourism programmes, 21.8% of respondents did not answer this question. Of those who responded, 54.0% say they are not interested in participating. No statistically significant differences were found between the family segments in terms of having previously participated in social tourism programmes (most of both segments reported not having participated), or regarding the question of whether they would like to participate in social tourism programmes.

Regarding the motives that families present for not having participated in social tourism programmes, there are differences between the two segments (Table 11.1). Respondents who belong to 'families with deprivation' tend to have more answers of 'lack of interest', 'financial reasons' and 'professional reasons', when compared with the total of respondents. 'Families without deprivation' refer above all to 'lack of knowledge' as a reason for not participating in social tourism programmes (Table 11.1).

11.4.2 Families who have not taken family holidays away from home in the past 4 years

In the NFH families surveyed, 47.5% of these families were materially deprived.

Almost all of the respondents (98.3%) who answered this question said they had never

Table 11.1. Differences between FH families in the reasons for not having participated in social tourism programmes.

Reasons for not having participated in social tourism programmes	Families without deprivation (*N* = 669)	Families with deprivation (*N* = 154)	All families (*N* = 823)	χ^2	Significance
Lack of opportunity	27.80%	30.90%	28.30%	27.261	0.001*
Didn't feel need	**11.20%**	1.00%	9.50%		
Lack of knowledge	24.20%	20.60%	23.60%		
Financial reasons	3.00%	6.20%	3.50%		
Professional reasons	0.60%	**4.10%**	1.20%		
Lack of information	2.50%	4.10%	2.80%		
Lack of time	5.90%	6.20%	6.00%		
Prefer holidays just with the family/with more freedom/privacy	**7.40%**	2.10%	6.50%		
Lack of interest	13.10%	**20.60%**	14.40%		
Other	4.20%	4.10%	4.20%		

*$P \leq 0.05$; ** $P \leq 0.1$; bold cells = adj. residuals ≥2.

participated in social tourism programmes (note that 12.6% of the total respondents did not answer this question). Only 1.7% of respondents who answered this question claim to have participated in social tourism programmes. This result is also a reflection of the absence of social tourism programmes for families with children in Portugal.

When questioned about why they have never participated in these types of programmes, the respondents mostly refer to (40.3%) the lack of knowledge about their existence as the main reason, with 0.7% of the respondents even mentioning the non-existence of these programmes (Fig. 11.2). The fact that 9.7% of respondents mention financial reasons as a reason for not

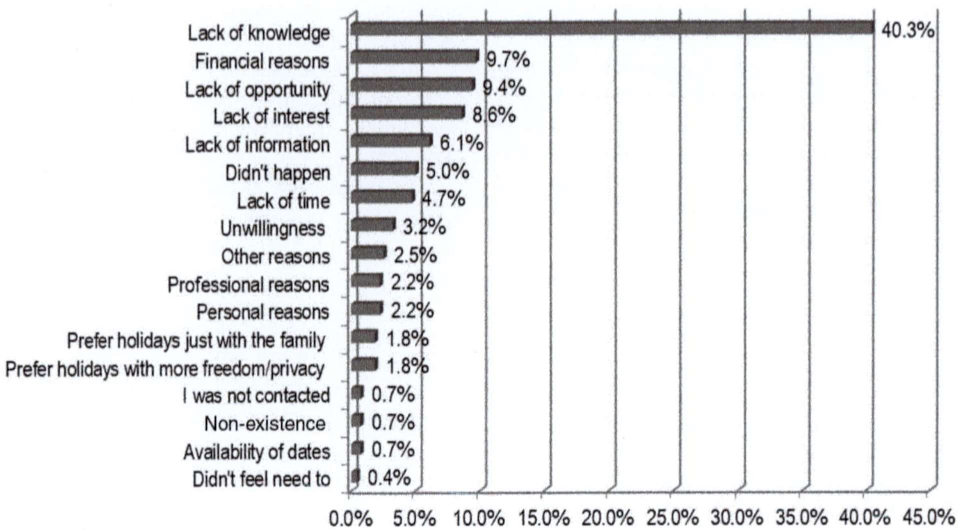

Fig. 11.2. NFH families' motives for not participating in social tourism programmes.

Table 11.2. Differences between NFH families reasons for not having participated in social tourism programmes.

	Families without deprivation (*N* = 169)	Families with deprivation (*N* = 153)	All families (*N* = 322)	χ^2	Significance
Would you like to participate in social tourism programmes?[a]	56.29%	**75.94%**	65.49%	12.080	0.001*

*$P \leq 0.05$; ** $P \leq 0.1$; bold cells = adj. residuals ≥2; [a] In the dichotomous variables, we only present the values for the 'Yes' category.

participating is noteworthy, even considering the type of trips that normally are included in these types of social tourism programmes.

Finally, 57.9% of respondents answered the question about whether they would like to participate in social tourism programmes and, of those who responded, 30.3% answered positively.

No statistically significant differences were found between 'families with deprivation' and 'families without deprivation', in terms of having previously participated in social tourism programmes, nor in relation to the reasons they present for not having done so.

However, when asked whether they would like to participate in social tourism programmes, respondents who belong to 'families with deprivation' tend to answer 'Yes' more often when compared with the total of respondents (Table 11.2). The fact that 'families with deprivation' tend to answer that they would like to participate in social tourism programmes, considering the definition of social tourism presented in the questionnaire, may be related to the fact that the 'families with deprivation' segment includes more individuals who attribute greater importance to 'financial' constraints as a reason for not taking a family holiday (Lima, 2015).

11.4.3 Comparison between FH and NFH families

Almost all of the respondents (more than 97% in both groups of families) claim never to have participated in social tourism programmes and hence there are no statistically significant differences between groups of families (between FH and NFH families nor between 'families with deprivation' and 'families without deprivation') on this issue.

There are statistically significant differences between FH and NFH families, as well as between 'families with deprivation' and 'families without deprivation' in terms of their willingness to participate in social tourism programmes and the reasons for not having participated (Table 11.3). NFH families and 'families with deprivation' demonstrated more interest in social tourism programmes, compared with the FH families and 'families without deprivation', respectively (Table 11.3).

When asked why they had never participated in this type of programme, the respondents who had not taken a family holiday in the past 4 years mostly refer, with more intensity than is the case for those who took a family holiday, to lack of knowledge about its existence as the main reason. FH families mentioned relatively more reasons related to 'lack of opportunity', 'I didn't have the need' and 'I don't appreciate it/lack of interest', while NFH families mentioned more 'financial' and 'unknown' reasons (Table 11.3).

Analysing the differences in the reasons presented for non-participation in social tourism programmes between 'families with deprivation' and 'families without deprivation', it is observed that 'families with deprivation' refer relatively more to financial reasons for this non-participation, while 'families without deprivation' refer relatively more to the fact that they did not have the need and prefer another type of holiday, just with the family, with more freedom and privacy (Table 11.3).

The two above-mentioned results in relation to the differences in the reasons for not participating in social tourism programmes are related

Table 11.3. Comparison of reasons for families not participating in social tourism programmes.

	NFH families (*N* = 340)	FH families (*N* = 825)	All families (*N* = 1165)	χ^2	Significant	Families without deprivation (*N* = 838)	Families with deprivation (*N* = 307)	All families (*N* = 1165)	χ^2	Significant
Would you like to participate in social tourism programmes?[a]	**65.67%**	46.05%	52.28%	31.595	0.000*	48.08%	**62.40%**	51.94%	14.999	0.000*
Motives for not having participated in social tourism programmes										
Lack of opportunity	14.39%	**28.42%**	23.82%	98.294	0.000*	24.84%	21.24%	23.86%	43.772	0.000*
Didn't feel need	0.00%	**9.47%**	6.37%			**8.72%**	0.44%	6.47%		
Lack of knowledge	**41.01%**	23.51%	29.25%			27.96%	34.07%	29.62%		
Financial reasons	**9.71%**	3.51%	5.54%			3.45%	**9.73%**	5.16%		
Professional reasons	0.00%	1.23%	0.83%			0.49%	1.77%	0.84%		
Lack of information	**6.12%**	2.81%	3.89%			3.62%	4.42%	3.84%		
Lack of time	7.91%	5.96%	6.60%			6.41%	6.19%	6.35%		
Prefer holidays just with the family/with more freedom/privacy	3.60%	6.49%	5.54%			**6.74%**	2.21%	5.52%		
Lack of interest	8.63%	**14.39%**	12.50%			12.66%	12.83%	12.71%		
Other	**8.63%**	4.21%	5.66%			5.10%	7.08%	5.64%		

*$P \leq 0.05$; ** $P \leq 0.1$; bold cells = adj. residuals ≥2; [a] In the dichotomous variables, we only present the values for the 'Yes' category.

to the fact that there are differences between FH and NFH families in terms of the sociodemographic profile. The NFH segment includes more families that seem to live in contexts of poverty and conducive to social exclusion, generally contexts with scarcer financial resources and also with less access to information, such as 'families with deprivation'. Thus, this more unfavourable life context of NFH families justifies why they indicate 'financial reasons' and 'lack of knowledge' as reasons for not having participated. On the other hand, more favourable contexts in which the FH families (and the 'families without deprivation') live, justify the greater relative frequency of the indication of reasons related to 'lack of opportunity', 'I didn't have the need' and 'lack of interest' (Table 11.3).

11.4.4 Low-income families

The interviewees were all Portuguese, mostly representing traditional families (father, mother and biological children), four of whom were single-parent families (three divorced and one widow). One of the families, in addition to the parents and children, included the maternal grandmother and two families that had recently integrated a new element resulting from a second de facto union/marriage. In these last two cases, the interview focused on the family holidays of the mother and children (single-parent family situation), since the union situation was recent and no holiday situation with this new element had yet occurred.

In terms of the sociodemographic characteristics of the 15 adults interviewed, 12 of them were women, 12 had educational qualifications at levels below higher education (mostly 6 and 9 years of schooling) and corresponded to professional categories, that is, mostly professions included in the category 'personal service, protection service and sales workers', from CPP/2010 (INE, 2011). It is significant that two of the interviewees represent families with one parent in an unemployed situation and another interviewee represents a family in which both parents are in this situation.

The totality of the interviewed families (15) state that they were not aware of social tourism programmes for families and, when they mentioned having heard of social tourism programmes, they referred to the senior tourism programmes promoted by INATEL. Consequently, none of the respondents had participated in these types of programmes. This result is very similar to those obtained in the application of the questionnaires, referred to in the previous sections.

Respondents were asked whether they would like to participate in social tourism programmes for families with children if they came into existence, and almost all said they would like to, depending on the conditions offered. The two mothers who said they had no intention of participating justified it with the statement that 'No, I don't like it because it is a lot of confusion. One day when I can, it will be just them. I want to enjoy the little time I have only with them' (Mother 2) and 'No. My husband has a health condition and he is not very keen on leaving the house. I think that at this moment I don't feel the need to leave the house' (Mother 4).

Respondents who want to participate, depending on their conditions, always referred to the positive effects of family holidays and the fact that it is not possible for them to do those holidays in any other way, as justifications for having that desire.

The next step was to understand which characteristics the families considered important in making these programmes attractive. The biggest and main concern mentioned is the 'affordable price', by which they meant 'price that they can pay'. Associated with this issue, there was also a reference to the fact that it would be easier for families to have these programmes provide a system of staggered advance payments, since 'we would pay x amount every month and when the time came, we would have it there ready to go. Because money for diesel, to eat and drink, it's like here...' (Father 14).

Then, the characteristics of these programmes were mentioned in terms of destinations, time and activities. Most respondents say that the destination should be the beach and at an appropriate season, in terms of weather conditions and school holidays. As for the activities that they should include, they emphasize that they should have the possibility of choosing which activities they wanted to engage in or not, in order to avoid preventing the achievement of their goal of relaxation, family time and the absence of rigid schedules, as illustrated by the statement of one of the mothers: 'I don't think it's right that they put schedules for everything. I think it couldn't be with everything scheduled. I think they could present the things that are there and then we would decide if we want to go or not' (Mother 11).

Another aspect mentioned in relation to activities is the fact that the interviewed families consider that there should be a part of activities aimed at parents and children and activities only for the children, which, however, must guarantee security. In other words, the activities should allow family life, but also entertainment for the rest of the parents and the children, as systematized by a mother: 'Activities a little bit for everyone, that we all do together and, from time to time, separate to rest a little, especially from the little one who is more complicated and requires a lot of attention' (Mother 8). Parents who reported preferring only joint activities did so essentially because of insecurity and the fear they feel about going outside the usual environment with their children.

These results allow the conclusion that attractive programmes are seen as always allowing the family to choose the type of activities they want to engage in: interaction with a larger group if the family wishes to, also moments of interaction between parents and children, allowing a relaxed relationship within the family, but providing moments when parents can, if they want to and feel safe, rest 'from the work of the children'. A mother's statement summarizes this idea: 'There has to be flexibility to be independent, to do what we want' (Mother 6).

Another mother points out that these programmes could be aimed at promoting family harmony, even with the intervention of professionals to guide the parents' behaviour:

> One thing that should exist is having a teacher or psychologist who asks the children what they think the parents are doing wrong or right and then the parents in relation to the children. Because sometimes, even though we think we are doing the best, we are not. And if there was an opportunity to hear from the other party, we would see things differently. And outside the usual environment, I would feel more at ease. Because being in our midst, we already know people and already know we will be criticised. And if it is in an environment that we do not know, we may be more withdrawn, in one aspect, but in another we are more comfortable to open up, to do certain things that we would not have the courage to do here. A programme that would even contribute to the union between parents and children. (Mother 7)

Finally, some interviewees even made suggestions on how this type of programme could be implemented, in terms of funding:

> There should be companies that tell us: you have access to a house, water, electricity but other expenses are your own. More affordable prices. For example, a company that had apartments and told us that we have a 3-bedroom apartment for two weeks, with expenses on our own, and we paid x amount per month, we would become partners, we would pay x amount every month and when the time came, we would have it ready to go. (Father 14)

With some interviewees, namely those who expressed greater desire for independence in this type of programme and families who usually take holidays more regularly (once a year or every 2 years), it was also possible to perceive some discomfort with the term social tourism programmes. There was even one interviewee who said: 'the need for social tourism, or call it another name, anyway ... family tourism' (Mother 5). This alerts to the fact that for these families, participating in social tourism programmes for families should not make them feel different from other families, nor in any way differentiated because they receive that support, under penalty of not even having interest in participating or that this feeling possibly compromises the benefits that they could derive from that participation.

11.5 Guidelines for Social Tourism Programmes for Families: Maximizing Benefits and Attractiveness

The aim of this section is to present some reflection on guidelines that social tourism programmes for families should follow in order to combine the maximization of its benefits and its attractiveness as a tourism product. However, it is not intended to assume that the results obtained from this investigation are completely sufficient to achieve this objective. It is considered that this chapter presents results that allow some clues to be identified about guidelines for the creation/improvement of social tourism programmes. More research is necessary in this sense, which will confirm and expand the exploratory results presented here.

The survey results indicate that almost all respondents did not participate in social tourism programmes, largely because they were unaware of the existence of this type of programme. In addition, NFH families say that they want to participate in social tourism programmes relatively more compared with FH families, as well as with 'families with deprivation' in relation to 'families without deprivation'. The results of the interviews with low-income families also indicate that families are receptive to a type of programme aimed at families with children, which allows them to benefit from all the benefits they perceive from family tourism, highlighting the importance of these programmes being, above all, affordable. These results show a window of opportunity for the creation of social tourism programmes aimed at economically deprived families, a product not currently offered in the Portuguese market in an organized and systematic way. The results of the survey showing the importance of financial reasons and lack of knowledge as reasons for not participating in these types of social tourism programmes seem to reinforce the existence of this opportunity.

Regarding the characteristics that social tourism programmes for families must incorporate in order to be attractive to families, this information was obtained only through interviews. From these results, the following characteristics are identified:

1. The main concern mentioned by the interviewees is that the price of these social tourism programmes has to be effectively accessible, considering their budgets.
2. The idea that it would be easier for families to have the option of payments in advance instalments throughout the year.
3. The destination should be the beach (a destination that contrasts with their usual environment – mountain region).
4. The season should be appropriate, considering the weather conditions and school holiday periods.
5. The activities to be included should always be optional and varied, providing interaction with a larger group if the family wishes to, interaction between parents and children, activities only for children (where the safety of these children is seen as a central concern) and activities only for adults, and 'free' moments (these indications vary according to the characteristics of each family).

Regarding the activities to be included in these social tourism programmes, another aspect that was mentioned was the possibility that these programmes could even integrate activities to promote family harmony, through the intervention of professionals who could 'guide' parents' behaviours. In the context of social tourism literature, the advice given by Minnaert (2006) and Minnaert *et al.* (2009) should be highlighted, regarding the continuing need to adapt the type of support to each family provided during the trips, an adaptation that will dictate the pursuit of the goals of changing behaviour and encouraging social inclusion.

The literature review supports the fact that the characteristics of tourist practices in which families participate will also determine the intensity and type of effects of this practice (Minnaert, 2006; Minnaert *et al.*, 2009; Moscardo, 2009; Alexander *et al.*, 2010; Dolnicar *et al.*, 2012; Lima, 2015). The results thus allow the conclusion that there is a need to adapt the type of holiday to each family (e.g.to meet the needs of different family members – children vs. adults, separation vs. socializing – or to meet the restrictions of children's school holiday periods) when the objective is to maximize the positive effects of family tourism.

Thus, the activities to be included will be a key feature of these social tourism programmes for families, both as a factor that will influence the type of experiences lived during the holidays and thus determinants of the effects of these holidays, but also as a determinant of the interest of economically deprived families in these social tourism programmes, recognizing the need for more research on this theme in order to identify what type of activities should be promoted for each type of family.

Families facing the 'new type of poverty' generally have different sociodemographic characteristics (e.g. higher educational levels or different professions from those associated with lower social status) and also different behaviours from 'traditional' low-income families. However, even though they do not yet have those patterns of organization and behaviour that are typical of families with 'traditional' low incomes, they are families that can easily adopt repetitive patterns of organization/behaviour as adaptation to circumstances of poverty (Lewis, 1970).

Therefore, one of the guidelines for social tourism programmes for families should be the differentiation of the type of holiday according to the characteristics of each family (from sociodemographic characteristics of family members, such as the age of children, to characteristics of the socioeconomic environment in which they live), especially in terms of the activities offered during the holidays and the specialized accompaniment that each family must have during those holidays. The specificities of each family and the concept of 'new type of poverty' must be considered. This differentiation requires at least two practical implications to which the agents that promote this type of programmes must pay attention:

1. The need to have qualified human resources from different scientific areas (e.g. psychology, sociology, economics or tourism), working on defining the components of the social tourism programmes.
2. The need to communicate with different social agents (e.g. social security, schools or other associations that exist and work with families) in the place of residence of each family in order to obtain real information about the characteristics of the family and thus effectively adapt the social tourism programmes to these families' characteristics, seeking to maximize the benefits of the social tourism programmes.

Another characteristic that seems essential in these social tourism programmes, which is more implicit in the interviews, has to do with the importance of these programmes not causing discomfort in families by identifying them as a 'needy' segment. This means the programmes must be designed so that the participating families do not feel different from other families, nor in any way differentiated, as they are receiving that support. Thus, the marketing strategy, especially the communication strategy, of these programmes must be carried out with particular care. An example of implementing this caution could be the communication of these programmes guided by the objective of promoting family well-being and/or developing programmes open to all types of families with children, with emphasis on the aspect of being accessible to all in the communication carried out. This opening up of programmes to all types of families, however, implies other challenges, such as:

1. The need to define, as a participation criterion, the payment of different prices according to income levels and/or the number of children in the family, as already happens internationally in some social tourism programmes.
2. The need to provide training for the agents involved in the process of accompanying families during the trip and for the supply agents of the destination.
3. Ensuring there are mechanisms to reduce the potential culture shock existing between different participating families, given the scope of the concept of 'low-income family'. These mechanisms imply specific caution from the programme organizers in terms of in-depth knowledge and understanding of specific characteristics associated with some of these families, for example, the 'new poor' or cultural and ethnic minorities.

Notwithstanding the aforementioned indications and findings from the literature and the empirical study, some challenges arise in the implementation of these social tourism programmes for families, which must be considered in their design. Current national budget constraints indicate that targeting public funds in this area may not be feasible in the short term, as there is a decrease in state funds for social policies that integrate social tourism programmes. However, these periods of greater economic austerity, such as the current scenario, are associated with a decrease in disposable household income, which contributes to an increase in social exclusion and, consequently, to an increase in the need for these programmes. This reality is yet another factor that gives rise to the need to develop social policies that provide mechanisms for social innovation, such as access for families to tourism. In this sense, a number of challenges arise for the implementation of these types of programmes.

In this context, it is imperative to find new sources of financing, appealing, for example, to the social responsibility of companies in the tourism area (and other sectors) to participate in this form of promoting social innovation in the fight against social exclusion. The opportunity represented by the priorities defined for financing with European structural funds until 2030 (linked with the achievement of the Sustainable Development Goals identified by the UN), and the creation of the 'Portugal Social Innovation' initiative can be important incentives, as they define social problems in general and combat poverty, social inclusion and human capital in particular, as priority lines of action. Non-profit organizations can play a decisive role, acting as intermediaries between public agents (national, regional and local), companies (possible sponsors of these initiatives, in the context of their social responsibility) and families, developing programmes adapted to the needs of each group of families and sensitizing economic agents to this measure to promote 'Tourism for All'.

Since there is evidence for the positive effects of tourism on the health of individuals and the consequent potential effect on health expenditure (Hall and Brown, 2006; INATEL, 2009a, b; Alexander *et al.*, 2010; Kuoni and Nuffield Health, 2013), another possible solution is coordinating social policy in the field of tourism with health policy, allocating some funds from the health area to support these initiatives. In the same way, it is considered that the existence of greater coordination between research and the bodies responsible for decision-making and the elaboration of intervention plans in this field are of paramount importance. The development of these types of programmes must pass through a study of the theories, foundations and practices related to social economy (Namorado, 2004) and even of solidarity economy (Dacheux and Goujon, 2011), where central values of cooperation, both formal and informal, and the concertation of objectives, lead to effective actions to promote the improvement of the living conditions of individuals. In this case, this particularly implies families that will participate in these programmes, but without forgetting the collective benefits that can result from the creation of third-sector organizations.

The importance of another guideline has to be highlighted: the need for an effective, cautious and gradual development process, especially since these social tourism programmes are considered as a product associated with a high level of innovation in the Portuguese context. This means that the creation of different offers of social tourism programmes for families must be implemented with an associated process of continuous monitoring, especially in terms of the reactions of families and the effects of social tourism programmes on these families, allowing adaptation and improvement of programmes to the identified circumstances.

These are some guidelines that are considered pertinent in the development of social tourism programmes for families, designed to maximize the return of the resources allocated by society to their implementation through the individual and collective benefits they originate (Fig. 11.3.). However, it is important not to forget that these guidelines should be seen as complementary to future studies that deepen this theme, since the results presented here are exploratory in nature.

11.6 Conclusions

Social tourism appears in response to concerns about social exclusion, providing individuals excluded from society with the opportunity to participate in tourism, a dimension that characterizes today's society and the quality of life. Social tourism may be considered, then, as an attempt to achieve equal access to tourism, inevitably integrating objectives related to combating social exclusion and poverty and corresponding to the objectives of social policies for inclusion and promotion of quality of life, through changing social exclusion behaviours. However, the integration of tourism programmes targeting disadvantaged social groups in social policies is a possible measure that has already aroused the interest of several entities, but has not yet received widespread and effective recognition and investment.

Sousa *et al.* (2007, p. 86) corroborate the need to use different approaches to combat social

Fig. 11.3. Systematization of the proposed guidelines for social tourism programmes for low-income families.

exclusion, stating that current social problems require measures that really promote the inclusion/participation of people in society, enabling its beneficiaries to have a positive and active role in society.

In this investigation, this opinion is shared, considering that policies that allow the development of behaviours that break the cycle of poverty and combat social exclusion in economically deprived families are the most appropriate nowadays. It was also argued that this type of policy may find an effective instrument to achieve its objectives in promoting and supporting tourism, as tourism is an activity that, when planned, can foster a strong connection/cooperation between society's public and private agents, promoting efficiency in that society.

The literature, although still insufficient, seems to indicate that incentives to tourism for disadvantaged groups, namely low-income families, can facilitate the social inclusion of these families, encouraging social cohesion and solidarity. This effect comes from the potential of tourism to provide visitors with a basis to activate their process of change, and it is up to planners and agents involved in the definition of social policies to develop mechanisms that enable low-income families to take advantage of this opportunity for change, maximizing the returns of policies to society. In this context, Minnaert *et al.* (2009) advise the need to adapt the type of support offered to each family during the trips, which will dictate the achievement of the goals of changing behaviour and social inclusion of these initiatives.

The results of the present study show that there are two main reasons for the respondents not participating in social tourism programmes: (i) the families were unaware of the existence of such programmes; and (ii) the non-existent supply of social tourism programmes for families in the Portuguese market. It is also clear that the surveyed families have an interest in participating in social tourism programmes, this interest being more evident in families with low income and families who have not had holidays in the past 4 years. Also, the families are receptive to a type of programme that permits them to benefit from all the positive aspects they associate with a family holiday, and these programmes should above all be affordable and allow the family to be independent at the destination. Despite the benefits that the families associate with participating in family tourism, the research identified a set of challenges faced by social tourism initiatives targeted to this segment in the future.

Results point to the prerequisite for the social tourism programmes for families to consider four types of needs that can contribute to their success, in terms of the objectives of maximizing the benefits that they can produce for families and for society, and of attractiveness for families: (i) being financially sustainable, both for the promoter and for families; (ii) being able to adapt their offer to each type of family; (iii) monitoring over time; and (iv) being able to adapt their marketing strategy (particularly communication) to different types of low-income families.

Finally, it is considered important to highlight some limitations that imply that the presented results be interpreted with caution. The small size of the sample of the qualitative study, the fact that this sample includes more mothers than fathers and the fact that entities that offer social tourism programmes (although not specifically for families) and that work directly with low-income families in Portugal were not included in this study are the main limitations to the clues given for the definition of social tourism programme guidelines for families. These limitations reinforce the need to conduct future studies that specifically address this issue.

Note

[1] The material deprivation rate considered for this study included nine items: ability to face unexpected expenses, ability to pay for 1 week's annual holiday away from home, existence of arrears, capacity to have a meal with meat, chicken or fish every second day, capacity to keep home adequately warm, possession of a washing machine, a colour TV, a telephone or a personal car (Wolff, 2009).

References

Alexander, Z., Bakir, A. and Wickens, E. (2010) An investigation into the impact of vacation travel on the tourist. *International Journal of Tourism Research* 12(5), 574–590. DOI: 10.1002/jtr.777.

Clark, A. E. (2003) Unemployment as a social norm: psychological evidence from panel data. *Journal of Labor Economics* 21(2), 323–351. DOI: 10.1086/345560.

Dacheux, E. and Goujon, D. (2011) The solidarity economy: an alternative development strategy? *International Social Science Journal* 62(203–204), 205–215. DOI: 10.1111/j.1468-2451.2011.01804.x.

Dolnicar, S., Yanamandram, V. and Cliff, K. (2012) The contribution of vacations to quality of life. *Annals of Tourism Research* 39(1), 59–83. DOI: 10.1016/j.annals.2011.04.015.

EC – European Commission (2010) Calypso study – compendium of good practices. European Commission dG enterprise and industry. Available at: file:///C:/Users/Joana/Downloads/calypso_compendium_of_good_practices_en.pdf (accessed 21 October 2020).

EESC – European Economic and Social Committee (2006) *Opinion of the European Economic and Social committee on social tourism in Europe, E.E.a.S. Committee*. Editor. EESC, Brussels.

ETB – English Tourist Board (1976) *Holidays: The social need*. S.T.S. Group, London.

Eurostat (2018) Glossary: material deprivation. Available at: https://ec.europa.eu/eurostat/statistics-explained/index.php/Glossary:Material_deprivation (accessed 19 October 2020).

Gray, M., Edwards, B., Hayes, A. and Baxter, J. (2009) The impacts of recessions on families. *Family Matters* 83, 7–14.

Hall, D. and Brown, F. (2006) *Tourism and Welfare: Ethics, Responsibility and Sustained Well-being*. CAB International, Wallingford, UK.

ILO – International Labour Organization (2020) ILO monitor 1st edition: COVID-19 and the world of work: impact and policy responses. Available at: https://www.ilo.org/wcmsp5/groups/public/-dgreports/-dcomm/documents/briefingnote/wcms_738753.pdf (accessed 19 October 2020).

INATEL (2009a) *Estudo de impacto sócio-económico do Programa Turismo Sénior de 2001 a 2005. [Study on the socioeconomic impact of the Senior Tourism Programme since 2001 to 2005]*. University of Aveiro (not published).

INATEL (2009b) *Estudo de impacto sócio-económico do Programa Saúde e Termalismo Sénior de 1997 a 2007. [Study on the socioeconomic impact of the Senior Health and Thermalism Tourism Programme since 1997 to 2007]*. University of Aveiro (not published).

INE – Instituto Nacional de Estatística (2011) *Classificação Portuguesa das Profissões de 2010. [National Occupation Classification 2010]*. Instituto Nacional de Estatística, I.P, Lisbon.

Kuoni and Nuffield Health (2013) The holiday health report 2013 – revealed: how holidays help you live longer. Available at: www.nhealth.co.uk./holiday-health-experiemnt (accessed February 2013).

Lehto, X.Y., Choi, S., Lin, Y.-C. and MacDermid, S.M. (2009) Vacation and family functioning. *Annals of Tourism Research* 36(3), 459–479. DOI: 10.1016/j.annals.2009.04.003.

Levitas, R. (2004) Let's hear it for Humpty: Social exclusion, the third way and cultural capital. *Cultural Trends 13(2, n)* 50, 41–56.

Lewis, O. (1970) *Os Filhos de Sanchez*. Moraes Editores, Lisbon.

Lima, J. (2015) Turismo em Família: A importância do turismo para famílias economicamente carenciadas. [Family tourism: the importance of tourism for low-income families.]. PhD thesis. University of Aveiro, Aveiro, Portugal.

Lima, J., Eusébio, C. and Amorim, C. (2012) Effects of holidays on tourists: a review and directions for future research. *RTD/Journal of Tourism and Development* 17/18, 265–277.

Lima, J., Eusébio, C. and Amorim, C. (2018) Impacts of family tourism on the families' quality of life – Differences according to the family economic profile. In: Campón-Cerro, A.M. and Hernández-Mogollón, J.M. (eds) *Best Practices in Hospitality and Tourism Marketing and Management: A Quality of Life Perspective*. Springer, New York, pp. 275–296.

McCabe, S., Joldersma, T. and Li, C. (2010) Understanding the benefits of social tourism: linking participation to subjective well-being and quality of life. *International Journal of Tourism Research* 12(6), 761–773. DOI: 10.1002/jtr.791.

McCracken, G. (1988) *The Long Interview*. Sage, Thousand Oaks, California.

Minnaert, L. (2006) Social tourism: a potential policy to reduce social exclusion? the effects of visitor-related social tourism for low income groups on personal and family development. PhD thesis. University of Westminster, London.

Minnaert, L., Maitland, R. and Miller, G. (2009) Tourism and social policy – The value of social tourism. *Annals of Tourism Research* 36(2), 316–334.

Minnaert, L., Maitland, R. and Miller, G. (2011) What is social tourism? *Current Issues in Tourism* 14(5), 403–415. DOI: 10.1080/13683500.2011.568051.

Moscardo, G. (2009) Tourism and quality of life: Towards a more critical approach. *Tourism and Hospitality Research* 9(2), 159–170. DOI: 10.1057/thr.2009.6.

Namorado, R. (2004) *A Economia Social – Uma constelação de esperanças. [The social economy – A constelation of hopes]*. Oficina do Centro de Estudos Sociais – Faculdade de Economia da Universidade de Coimbra, Coimbra, p. 213.

Patton, M.Q. (1990) *Qualitative Research and Evaluation Methods*. Sage, Thousand Oaks, California.

Paul, K.I. and Moser, K. (2009) Unemployment impairs mental health: Meta-analyses. *Journal of Vocational Behavior* 74(3), 264–282. DOI: 10.1016/j.jvb.2009.01.001.

Rede Europeia Anti-Pobreza/Portugal – REAPN (2008) Reacção ao Plano Nacional de Acção para a Inclusão – PNAI 2008–2011. [Reaction to the National Action Plan for Inclusion 2008–2011.]. Available at: www.reapn.org/documentos_visualizar.php?ID=141 (accessed 30 January 2010).

Smith, V. and Hughes, H. (1999) Disadvantaged families and the meaning of the holiday. *International Journal of Tourism Research* 1(2), 123–133. DOI: 10.1002/(SICI)1522-1970(199903/04)1:2<123::AID-JTR146>3.0.CO;2-R.

Sousa, L., Hespanha, P., Rodrigues, S. and Grilo, P. (2007) *Famílias Pobres: Desafios Intervenção. [Poor families: Challenges for intervention.]*. Climepsi editores, Lisbon.

Standard & Poor's (2014) Economic research: how increasing income inequality is dampening U.S. economic growth, and possible ways to change the tide. Available at: https://www.globalcreditportal.com/ratingsdirect/renderArticle.do?articleId=1351366&SctArtId=255732&from=CM&nsl_code=

LIME&sourceObjectId=8741033&sourceRevId=1&fee_ind=N&exp_date=20240804-19:41:13 (accessed 5 August 2014).
Sumner, A., Roy, C. and Ortiz-Juarez, E. (2020) Estimates of the impact of COVID-19 on global poverty. wider working paper 2020/43, United nations university world Institute for development economics research (UNU-Wider). Available at: https://www.wider.unu.edu/sites/default/files/Publications/Working-paper/PDF/wp2020-43.pdf (accessed 19 October 2020).
UN – United Nations (2020) Sustainable development goals. Available at: https://www.un.org/sustainabledevelopment/development-agenda-retired/ (accessed 19 October 2020).
Wolff, P. (2009) Population and social conditions, in Eurostat: Statistics in focus 46/*2009*. Available at: http://epp.eurostat.ec.europa.eu/cache/ITY_OFFPUB/KS-SF-09-046/EN/KS-SF-09-046-EN.PDF (accessed May 2010).
Yin, R.K. (2011) *Qualitative Research from Start to Finish*. The Guilford Press, New York.

12 What, and Who, Is Social Tourism? The Roles of the Public, Private and Voluntary Sector in Social Tourism Provision in Flanders, Belgium

Lynn Minnaert*
New York University, New York, USA

12.1 Introduction

In many countries in mainland Europe, social tourism is reliant on some form of stimulus funding. This has often been provided by the state or the public sector, and yet the role of the public sector in the provision of opportunities for tourism participation is on the decrease (McCabe *et al.*, 2011). Prior and current economic downturns have led to public sector spending coming under increased scrutiny, and social tourism funding is moving away from investment in physical resources and towards stimulus for collaborative or partnership programmes, in a pluralistic financing model. One of the major challenges for social tourism in many European countries is the growing requirement to become less dependent on state funding and to become financially self-sustaining, requiring traditional social tourism providers to reposition themselves to stay competitive.

The organizational structure of social tourism is changing in response to a dynamic tourism sector, which poses challenges as well as opportunities. In several countries, the social tourism infrastructure has been privatized over recent years, and some social tourism accommodation centres now attract a very different clientele from that which it was originally intended for (McCabe *et al.*, 2011). In some cases it is not clear whether social tourism facilities actually benefit social tourists. There are growing similarities between, and indeed somewhat of a convergence of, social and commercial tourism: in many countries social tourism organizations now have to compete with private competitors. Social tourism accommodation today is often of relatively high quality, well-located and may meet demanding environmental sustainability criteria – from that perspective, they are often hard to distinguish from commercial competitors. This causes some to question whether there is still a need for the traditional social tourism sector: it could be argued that the availability of budget accommodation on a mass European scale is more successful in reaching low-income citizens. Indeed, in many social tourism systems (e.g. Holiday Participation Centre in Flanders, IMSERSO in Spain and the Cheques Vacances in France) private-sector partners play a large role in social tourism provision. This begs the question: how exactly does one classify social tourism initiatives? Which conditions do providers need to meet to be classed as 'social tourism', and can

*l.minnaert@nyu.edu

 Social Tourism: Global Challenges and Approaches (eds J. Lima and C. Eusébio)
DOI: 10.1079/9781789241211.0013

profit-seeking private enterprises also be classed under this label if these conditions are met? This chapter will explore these questions by examining the case study of Flanders. Flanders is the northern, Dutch-speaking region of Belgium, where tourism is under the authority of the regional government.

12.2 Flanders, Belgium: Social Tourism in Transition

A pivotal point in the development of social tourism, and indeed popular tourism in Belgium, was the 1936 Holiday with Pay Convention, initiated by the International Labour Office in Geneva. Article 2.1 of this convention states that 'every person to whom this convention applies shall be entitled after one year of continuous service to an annual holiday with pay of at least six working days'. The Holiday with Pay Convention is generally considered as the starting point for social tourism in Europe (Lanquar and Raynouard, 1986; Chauvin, 2002).

The years following the war were the heyday for social tourism. The period between 1950 and 1980 is described in French as the 'trente glorieuzes', the glorious thirty. Traditional social tourism was provided by labour unions, mutual companies and other non-profits such as religious organizations. It was based around the holiday centre: a product that was new, desirable and affordable and helped towards a democratization of holiday making. Traditionally, holiday centres offered a stay in full board with all entertainment included. Guests stayed in rather basic accommodation at low rates, and often helped with the daily chores. Management was often bureaucratic, and the operation of the holiday centres relied heavily on public subsidies. Over time, holiday centres evolved according to the needs of their public: many switched from full board to half board, the visitors had more freedom when choosing their activities, and help with the chores was no longer required (Chauvin, 2002, p. 67).

Growing post-war prosperity in Belgium and the emergence of mass tourism, however, started posing challenges to the traditional social tourism model. As commercial and international tourism options became more affordable, traditional holiday centres started losing their appeal, and the public funding structure changed accordingly. The Royal Decree of 1980 presented a shift in subsidy allocation towards youth hostels and camp grounds. In 2003, this decree was replaced by the Flemish 'Tourism for All' decree, whereby 'Tourism for All' is defined as 'a *non-commercial* form of tourism and/or recreation, which is equal in value to other forms of tourism and of recreation' (Tourism Flanders, 2018). More specifically, the decree stipulates that social tourism (referred to as 'Tourism for All'):

1. Is aimed at and has a mediating function towards all who are prohibited from fully participating in holidays away from home, e.g. families, youth, people with disabilities, the elderly and one-parent families, whatever their age, health or their economical/social/cultural background may be.
2. Aims for *non-commercial* participation in tourism, concentrating not only on pure relaxation, but also on recreation for the family, the group or the individual, with the purpose of improving physical, psychological, social and cultural well-being. This includes day trips and longer stays.
3. Works on the basis of a socially and economically acceptable pricing policy, stimulating specific target groups such as youth, families, people with disabilities, low-income groups and one-parent families.

The 'Tourism for All' decree presented a policy shift in that the target users of social tourism were more clearly defined as those who, despite the democratization of tourism, were still unable to participate in holidays. Social tourism in this decree is contrasted with private tourism businesses, and the 'non-commercial' nature of social tourism is heavily emphasized.

Tourism Flanders, the regional destination marketing organization, was founded in 1984 under the name 'Commissary General for Tourism'. Social tourism was part of its portfolio, and in 2001, the 'Holiday Participation Centre' was launched to focus specifically on the needs of Flemish residents who are unable to travel. The Holiday Participation Centre acts as an intermediary between supply and demand for social tourism. On the supply side, it works with traditional social tourism providers as well as private businesses – it centralizes information about discounted travel options for low-income travellers. On the demand sides, it liaises with social organizations and non-profits in Flanders that support the target users. Some of these social organizations belong to the public sector (social services departments, welfare offices), whereas others belong to the voluntary sector (charities, not-for-profit organizations). The social organizations ensure that the reduced rates offered are accessible only to the users they are intended for, so the system is not subject to abuse (Minnaert, 2011). Beneficiaries of the initiative have access to reduced rates for accommodation, attractions, events and group holidays (e.g. summer camps for children) within the region. In 2017, the Holiday Participation Centre reached 150,000 beneficiaries (www.vakantieparticipatie.be/content/numbers).

If we consider the Holiday Participation Centre as the facilitator among a network of suppliers and social organizations, we see that the majority of stakeholders are in the public and not-for-profit sectors. This applies to all social organizations, and many supplier partners, such as youth hostels and traditional social tourism holiday centres. The Holiday Participation Centre, for example, has certified 587 accommodation providers as social tourism providers (representing over 45,000 beds), which allows these to apply for social tourism subsidies. The large majority of these providers operate in youth tourism (Tourism Flanders, 2018). However, the Tourism Participation Centre also has stakeholders in its network from the private sector. It offers discounted stays at bed and breakfasts and chain hotels such as Ibis, and discounted admittance to private attractions, museums and events.

This begs the question of the extent to which the *non-commercial* nature of partners is a required feature of social tourism initiatives. The 'Tourism for All' decree was rooted in a traditional view of social tourism that drew a stark dividing line between social and commercial tourism. In the wake of the financial crisis of 2008 and the austerity measures many European governments implemented, there is a tendency to reduce public funding for social welfare initiatives. In a time of mass tourism, low-cost airlines and cheap package holidays moreover, many of the original social tourism establishments find it hard to compete with the commercial sector. The traditional social tourism sector thus faces a dilemma: because of the reduction in public funding, the cost of social tourism provision increases, while at the same time, owing to economies of scale and an increasingly integrated tourism sector, the prices of commercial tourism often decrease. This has led to some interesting anomalies, whereby it is sometimes cheaper to fly for a week all-inclusive to an international destination than to spend a week in domestic social tourism accommodation (McCabe *et al.*, 2012). As the tourism sector has professionalized, accessibility and corporate social responsibility (CSR) are a key focus for many private enterprises – inclusivity and a welcoming environment for all have become priorities in public discourse and the business environment, and are no longer relegated to the margins of tourism operations. The following section will introduce the concept of CSR and its importance in private-sector business strategies.

12.3 CSR and Private-Sector Contributions to Social Tourism

A range of definitions for CSR have been proposed, and little consensus between authors exists on their relative value. CSR can be seen as 'the configuration of the principles of social responsibility, processes of social responsiveness, and policies, programmes and observable outcomes as they relate to the firm's societal relationships' (Wood in Maignan and Ralston, 2002, p. 497). Alternative definitions are 'actions that appear to further some social good, beyond the interests of the firm and that which is required by law' (McWilliams *et al.*, 2005, p. 3); and 'policies and activities that reach far beyond mandatory obligations such as economic responsibility (being profitable) and legal responsibility (obeying the legislation and adhering to regulations)' (Halme and Laurila, 2009, p. 327).

Halme and Laurila (2009) further suggest that there are three types of CSR initiatives. *CSR philanthropy* focuses on charity, sponsorship and employee voluntarism. *CSR integration* centres on conducting existing business operations more responsibly, and is aimed at primary stakeholders such as customers, employees and suppliers. *CSR innovation* places the emphasis on developing new business models for solving social and environmental problems – social enterprises are examples of this type. This typology shows that CSR encompasses more than just community activities such as donations to good causes and involvement in local development initiatives. Community activities, however, are probably 'the most obvious manifestation of the voluntary and discretionary elements that are often seen as central to CSR' (Crane *et al.*, 2008, p. 266).

The type of CSR that focuses on non-primary stakeholders (persons who are not shareholders, clients or suppliers) in the wider community is also referred to as corporate philanthropy, which can take 'the form of financial donations as well as contributions in terms of free products and employee time' (Crane *et al.*, 2008, p. 266), as well as gifts in kind, loans of equipment, the provision of facilities and professional advice, advertising in charity magazines and brochures, and 'non-commercial sponsorship' (Bennett, 1998). 'Corporate philanthropy differs from commercial sponsorship in that it (supposedly) involves donations made without any expectation of direct commercial reward' (Bennett, 1998, p. 459).

Financial donations of companies are generally seen as being in decline: they have been criticized for being 'too paternal', for being a public relations device, for adding little of value to the cause and because they can be seen as providing little real benefit to the firm. In recent years, many companies have adopted a more strategic approach to corporate philanthropy, typically going under a label of 'strategic philanthropy', or 'corporate community investment' (Crane *et al.*, 2008, p. 266). Hess *et al.* (2008, p. 270) agree that 'many of these new corporate social initiatives are taking on aspects more commonly associated with corporate strategy than community relations: they are grounded in the core competencies of the firm and are related to the firm's long-term strategy'. An example is UPS, a logistics company that offers free shipping of aid packages to war or disaster-stricken areas, rather than just donating money (Hess *et al.*, 2008).

Careful management of CSR and corporate philanthropy initiatives should ensure benefits for the business itself, as well as for the community it is aiming to support. Maignan and Ralston (2002) describe CSR as an instrument to help achieve the firm's performance objectives, in a way that conforms to stakeholder norms defining appropriate behaviour. CSR should be based on carefully chosen principles: central, enduring and distinctive values to the firm. As such it may be regarded as a component of corporate identity. CSR can also be used as an impression management tool employed to influence stakeholders' perceptions of the company. Some comparative studies indicate that 'long-lasting community involvement programmes are more likely to improve the image of the corporation than after-profit cash contributions' (Hess *et al.*, 2008, p. 273). Porter and Kramer (2008, p. 291) claim that engaging in corporate philanthropy can improve the competitive position of businesses or business clusters: 'Philanthropic investments by members of a cluster, either individually or collectively, can have a powerful effect on the cluster's competitiveness and the performance of all its constituent companies.' In addition, CSR can be seen as a win–win situation for businesses and governments: 'Governments expect this trend to advance social justice and to slow environmental degradation, while companies usually hope that it will allow them to retain their license-to-operate and minimise mandatory intervention by external parties' (Halme and Laurila, 2009, p. 325).

The discussion above highlights that CSR and corporate philanthropy are not necessarily completely non-commercial. Private enterprises often aim to combine their philanthropic/social efforts with strategic actions that support their business objectives. Some authors are critical of the potential of CSR initiatives to achieve significant benefits. Porter and Kramer (2008, p. 287) suggest that 'the majority of corporate contribution programmes are diffuse and unfocused'. In addition, commentators are sometimes sceptical about the motives of corporations, and indicate that the only objective for the firm may be to achieve positive PR: 'what passes for "strategic philanthropy" today is almost never truly strategic, and often isn't even effective as philanthropy. Increasingly, philanthropy is used as a form of public relations or advertising, promoting a company's image or brand through cause-related marketing or other high-profile sponsorships' (Porter and Kramer, 2008, p. 286). Despite these criticisms, 'there appears to be significant public support and high expectations for CSR programmes' (Hess *et al.*, 2008, p. 275). This support could stem from 'the potential comparative advantage of business over governments or non-profits to provide assistance in solving certain social problems' (Hess *et al.*, 2008, p. 276). Indeed, in times of increased scrutiny over the use of public funds, and heightened demands for social initiatives to be financially self-sustaining, incorporating the private sector into social initiatives may become a more pressing demand. If this is the case, at what stage, if any, do private businesses who participate in social tourism initiatives 'become' social tourism?

So how exactly do we define what, and who, is social tourism today? Is social tourism still strictly seen as a non-commercial activity, or are there areas where commercial and non-commercial interests could intertwine? Is the cost of the tourism product or service the defining factor – in other words, is the cheapest product also by nature the most inclusive towards low-income groups in society? What factors determine whether we classify an initiative or product as social tourism? These research questions were examined in the study presented in the sections that follow.

12.4 Methodology

This chapter reports on a research project conducted with the support of Tourism Flanders. A series of nine focus groups (91 respondents) was convened with a range of tourism stakeholders: the respondents represented the public, private and voluntary sector, and included both those involved in social tourism provision (under the traditional definition) and those who are not. The focus groups were hosted in five different locations in the region, and participants were selected from the Holiday Participation Centre network in each location. The focus groups made use of interactive activities to guide the discussion: respondents were asked to class tourism scenarios as 'social tourism' or 'not social tourism', and were asked to place social tourism characteristics (e.g. price, social position of participant, level of sustainability) in order of importance. The aim of the focus groups was to discover what makes an initiative 'social tourism', and to define the concept not only by what it is (or should be), but also by what it is not (or should not be).

In the first activity, respondents were divided into smaller groups (three to four participants per group), and were presented with ten statements on coloured index cards. Each group was asked to divide the cards into two groups: one group for statements they felt could be classified as social tourism, one group for statements they felt were not social tourism. The statements were specifically chosen to represent traditional social tourism organizations, not-for-profits, low-cost travel options, CSR initiatives and business practices that are inclusive. The statements the respondents were presented with were:

- EasyJet and Ryanair make air travel affordable for everyone.
- A holiday centre operated by a union.
- A hotel offers gluten-free meal options.
- Youth movements such as the Girl Scouts or the Boy Scouts host summer camps at low rates.
- A private investor builds a care hotel (for convalescence) and offers rooms at market rates.
- A non-profit fundraises to pay for a weekend away for teenage mothers.
- A private tour operator donates a percentage of its profits to charity.
- The zoo offers heavily discounted tickets to low-income visitors via the Holiday Participation Centre.
- A theme park has special discounts for families with three or more children.
- A business hotel that employs people who were previously incarcerated.

In the second activity, respondents were presented with a list of characteristics that may (or may not) be attributed to social tourism providers. Each characteristic was listed on a white index card. Respondents were asked to rank these characteristics from most to least defining of social tourism. They were also provided with blank index cards in case they wanted to add additional characteristics to the list. The characteristics that were presented were:

- Initiatives that specifically aim to reduce travel barriers.
- Strong social mission.

- Low rates.
- Income level of the guests/visitors.
- Operated by a not-for-profit, union or mutual company.
- Inclusive hiring strategies.
- Linkages with local communities.
- Environmental sustainability.

The focus groups were audio recorded and transcribed. The data were analysed on the basis of a thematic analysis method. The findings of the study are presented in the following section.

12.5 Findings

In this section, we will review the findings from the eight focus groups. The findings will be split into two subsections. The first subsection will discuss the activity wherein respondents were presented with ten statements, and were asked whether in their opinion these statements were examples of social tourism or not. The second subsection will discuss the activity wherein respondents were presented with eight characteristics of social tourism, and they were asked to rank these from most to least applicable. Respondents were offered blank cards to add additional characteristics.

12.5.1 Statement activity

This activity generated intense discussion, and often there was no consensus in the room, nor in the smaller activity group, on whether or not a statement could be classified as social tourism or not.

The statements that were classified as social tourism most often were:

- The zoo offers heavily discounted tickets to low-income visitors via the Holiday Participation Centre.
- Youth movements such as the Girl Scouts or the Boy Scouts host summer camps at low rates.

Of all the statements, only 'The zoo offers heavily discounted tickets to low-income visitors via the Holiday Participation Centre' was classified as social tourism by all respondents. The fact that the initiative was coordinated by the Holiday Participation Centre was often named as a deciding factor in this classification. The fact that the zoo is a long-term partner of the Holiday Participation Centre, and that the discount provided to beneficiaries of the Centre is substantial (approximately 75%) are added arguments in favour of designating the initiative as social tourism.

Summer camps organized by youth movements, such as the Boy Scouts and Girl Scouts, were also classified as social tourism by almost all groups. The reasons were that these organizations support and connect their members throughout the year, and that the camps have a clear social mission. A few respondents however objected, highlighting that the membership of youth movements tends to be dominated by the middle and upper middle classes: these respondents felt that the youth movements did not achieve a large enough membership in socially disadvantaged circles.

The statements that were classified as social tourism least often were:

- EasyJet and Ryanair make air travel affordable for everyone.
- A hotel offers gluten-free meal options.
- A private tour operator donates a percentage of its profits to charity.
- A business hotel that employs people who were previously incarcerated.

Several of the above statements sparked debate, with some respondents offering counter-arguments to the majority view that these are not examples of social tourism.

For the example of low-cost carriers such as easyJet and Ryanair, the lower cost of the travel product was the main reason for debate. The low fares these carries offer were highlighted as enabling people from different background levels to travel internationally and by air – they were noted as effective in removing certain financial barriers. One respondent commented 'If a company makes something accessible to all and offers it to all, is it possible to be more "social" than that?' The majority of the respondents, however, did not agree, and raised objections to seeing these carriers as representative of social tourism. One example was the objection that while easyJet and Ryanair offer cheap flights, there is no further support for low-income users, who may lack the travel experience to confidently organize the rest of the trip. Other objections were that these carriers do not have a social mission, at times they have a negative image as employers, and they offer only minimal services for travellers with disabilities. As one respondent put it: 'I don't think these companies ever specifically think about excluded groups – this is just a business model that works across a large cross-section of society'.

The statement about the hotel offering gluten-free meal options also caused extensive discussion: the practice was mentioned as inclusive by several respondents, who also commented that people who require gluten-free diets are still often underserved in many Flemish destinations. However, several objections were raised to classifying this as a social tourism practice. For some respondents, this element was not strictly tourism-related, and hence did not deserve the designation. Others felt that this was not an example of a 'social' initiative, bringing up comparable examples such as vegetarian meals, which are not usually seen as a distinctive part of any social tourism initiative. In most groups, the consensus was that this was an inclusive and commendable business practice, but not an example of social tourism.

The statement about the private tour operator that donates a part of its profits to charity was almost universally classified as not an example of social tourism. The consensus was that this was an example of CSR. Only one group classified the statement as a 'maybe', arguing that if the money raised supported a social tourism initiative, the term could still apply. However after discussing the example in more detail with other groups, who argued that the term should only apply to the social tourism organization that was the beneficiary of the donation, this group moved the statement to the 'not social tourism' list.

Finally, the statement about the business hotel employing people who were previously incarcerated was almost unanimously classified as not social tourism. Instead, respondents referred to it as an example of social entrepreneurship: highly commendable, but not social tourism as the beneficiary in this case is not the person travelling. One respondent commented:

> I believe we need to distinguish here between the goal and the means. In this example, a profit is generated by a tourism enterprise, and that profit supports inclusive employment practices. Tourism is the means, employment is the goal. It's a laudable initiative, but for me it isn't social tourism. For it to be social tourism, the travel experience itself needs to be the catalyst for the benefits achieved.

The four remaining statements generated the fiercest discussions, and in some cases, no consensus was reached as to whether the statement was an example of social tourism or not. For some statements, different focus groups reached a different consensus. These statements were:

- A private investor builds a care hotel (for convalescence) and offers rooms at market rates.
- A non-profit fundraises to pay for a weekend away for teenage mothers.
- A theme park has special discounts for families with three or more children.
- A holiday centre operated by a union.

The statement about the private investor building a care hotel for convalescing patients, offering rooms at market rates, was hotly debated. For some respondents, the fact that the hotel is targeted towards an underserved and often excluded group of travellers was a deciding factor in classifying it as social tourism. For others, however, the fact that the hotel charges market rates was seen as an impediment: unless public funding mechanisms were available, or unless the hotel would consider offering a percentage of rooms at discounted rates for low-income guests, they did not consider the statement to be an example of social tourism. Several respondents felt strongly that the mission of the hotel was by definition inclusive and therefore 'social', whereas for others that was not the deciding factor: they insisted that this was just an example of a private business capitalizing on a social problem or need. In three focus groups, this statement was classified as social tourism, in five it was not and in one no consensus was reached.

For the statement about the non-profit fundraising to pay for a weekend away for teenage mothers, the majority of groups concluded it was an example of social tourism if certain conditions were met. These respondents felt it was important to distinguish between philanthropy and social tourism. One respondent referred to the statements as potentially 'American style charity – a lovely initiative in itself, but not social tourism, as the initiative isn't part of a larger support mechanism'. Others who felt similarly felt that the question hinged on whether the money was just collected and donated to the group of teenage mothers, or whether the not-for-profit would organize and facilitate the trip. In the former case, they would not see it as social tourism; in the latter case, they would. Once shared with the other groups, this reasoning was accepted in all focus groups, and the example was listed as a 'maybe'.

The statement about the theme park offering special discounts for families with three or more children received the most diverse set of responses: in three focus groups it was classified as social tourism, in four it was not and in two no consensus could be reached. For those who classified the initiative as social tourism, the fact that discounts were offered to larger families indicated that a financial challenge was addressed, and that the theme park showed dedication to making its product available to all. Others, however, felt strongly that the number of children was not a good indicator of the financial means of the family: for example, they highlighted that a single parent with two children may face larger financial challenges than a high-earning family with four children. These respondents suggested that the criteria for the discount needed to be refined, and that the theme park was only acting out of commercial interests by trying to attract larger families. In one focus group, this sparked particular debate about the feasibility of establishing more detailed criteria: how would evidence be obtained that families met these? Two respondents felt strongly that this would be unfeasible, and that by having an initiative that was easily implemented on a large scale, social tourism beneficiaries would have a chance to benefit too – even if it means that some people will benefit who do not necessarily need to overcome a financial barrier.

Interestingly, one of the statements that referred to a traditional social tourism provider was also hotly contended. The statement about the holiday centre operated by a union resulted in a lack of consensus in five focus groups. Of the four remaining focus groups, two classified it as social tourism, two as not social tourism. This was in some way an unexpected finding, as unions and mutual companies are two of the founding bodies of large-scale social tourism in Flanders: most still operate a network of holiday centres across the region. So why did so many respondents feel reluctant to classify them as social tourism? One respondent summed it up as follows: 'while the unions are the founders of social tourism, they have now largely made themselves redundant'. Many respondents highlighted that rates at union properties may be discounted for union members, but that in many cases they are still high, as the properties have been upgraded over the years and now offer a more high-quality product. As one respondent put it:

> The unions originally represented the working classes, and people who were unable to afford other travel options. Now many union members have good salaries and can afford to go where they please. People who cannot afford to travel are less likely to be union members, as they may not be (or no longer be) in the workforce.

For many respondents, the holiday centres are mainly a membership benefit, and no longer aimed at encouraging travel participation of people who would otherwise be excluded from tourism. In several focus groups, however, the union-operated holiday centres also had strong supporters. For some, their historical value is a key argument to maintain the social tourism classification. One respondent commented: 'Call me old-fashioned, but unions are about values – connecting people, fighting for what's right. For me, they still have that added value over other lodging options.' Another respondent mentioned:

> There can be other barriers than just the financial challenge – it's also the welcome a guest receives. Union holiday centres may not always offer the lowest price, but they can offer great value nonetheless: they will make sure every guest feels at home, and their rates will often include food and activities. While they may not have the lowest price, their rates are lower than some comparable accommodations for what you get – that still makes it 'social tourism' for me.

12.5.2 Characteristics activity

After the statement activity had highlighted how challenging it could be to classify what is social tourism and what is not, respondents were asked to complete a second activity, where they were presented with eight characteristics that may or may not apply to the concept. They were asked to rank the characteristics from most applicable to least applicable. The characteristics respondents were provided with were:

- initiatives that specifically aim to reduce travel barriers;
- strong social mission;
- low rates;
- income level of the guests/visitors;
- operated by a not-for-profit, union or mutual company;
- inclusive hiring strategies;
- linkages with local communities;
- environmental sustainability.

The two characteristics that were most likely to be placed at the top of the list were:

- initiatives that specifically aim to reduce travel barriers;
- strong social mission.

The first one was chosen as the top characteristic by most groups: it highlights the appreciation that respondents had for targeted initiatives, specifically geared to social tourism beneficiaries. Many highlighted that this for them pointed to the distinction between inclusive and social tourism: while inclusive initiatives aim to improve the travel experience for everyone, social tourism initiatives specifically target travel barriers such as lack of money, lack of travel experience, health and access barriers. Several respondents emphasized that social tourism is rooted in a holistic approach, which considers the whole travel experience. One respondent gave this example:

> It isn't enough to offer a discounted entry fee to an attraction if our beneficiaries then have to fork out for expensive drinks for the family. For those in the lowest income brackets, a trip may involve going grocery shopping together, and preparing lunches together. Touches like that can make a world of difference: for those in extreme poverty, it can be hard to just blend in with the mainstream as they face so many specific challenges.

Another commented that:

> Some of the beneficiaries need a lot of help. They may not know where to take the bus, they may need to borrow a suitcase. A general tourism provider would never offer services like these.

However, not all groups agreed on how the reduction of travel barriers could be achieved. Only one group of social tourists need this much support, they argued, so there may be a role for mainstream tourism providers with a strong social mission. While this characteristic was highly ranked by many groups, not all respondents agreed. Some questioned whether having a social mission was little more than a PR initiative, such as this respondent:

> I didn't want the strong social mission to be at the top of the list. Nowadays, all companies have a mission, and often it is just talk. I often think it's all done for marketing purposes: 'look how great we are'. But then you see what they do in practice, and it's mostly window dressing. I think practice is more important than mission.

While many other respondents agreed with that statement, the majority still felt that a strong social mission – a well-thought-out intention to achieve social good – was a key characteristic of social tourism, with most groups having the characteristic ranked as most or second most important.

Two characteristics that were mostly placed towards the middle of the list were:

- low rates;
- income level of the guests/visitors.

Both characteristics refer to financial barriers, and while respondents found these important, they had some reservations in placing them at the top of the list. For many, low rates are important, but may not tell the whole story: the need for a holistic approach, discussed earlier in this section, made the respondents distinguish between price and value. Low-cost carriers, also discussed earlier, are an example of a provider offering great prices, but little support: this means that while affordability is often paramount in social tourism, it cannot be seen in isolation from the reduction of travel barriers and the social mission of the organization.

Similarly, the income level of the guests and visitors was discussed as an important monitoring tool to measure whether certain target groups had been reached, but respondents highlighted that it isn't always easy to track or record data on income in tourism settings. Also, they emphasized that attracting mixed audiences should be a goal of social tourism, to promote social inclusion and cohesion. One respondent phrased it this way:

> I believe inclusion has to be central – maybe it should even be on a separate index card. Social tourism should not be just tourism for the poor. However not all of us in the group agreed – some felt it is important to sometimes be exclusive, and to just serve people with a certain need.

The characteristic most hotly debated was:

- Operated by a not-for-profit, union or mutual company.

This characteristic led to strong differences of opinion. While the majority did not feel that this characteristic was important in defining social tourism, a number of respondents felt it was of paramount importance. One group mentioned: 'We don't feel this is a characteristic that still applies today. It's not at the heart of social tourism. For the beneficiary, it makes little difference whether the initiative is public or private.' However in several other focus groups, the issue was hotly debated. As an example, an exchange between three respondents is summarized here:

> Respondent 1: 'I think this is key to the discussion. Because I don't buy it: private corporations need to make a profit, so I don't see how we could expect any social tourism from them. A private business needs to be profitable; that is the bottom line. A government agency or not-for-profit also needs to balance its books to survive; that is obvious, but the end goal is different.'
>
> Respondent 2: 'If we hadn't had support from private companies, though, to help us achieve our goals, we would be nowhere today. We should change the narrative from "either – or" to "x as well as y". The choices don't have to be this polarised.'
>
> Respondent 3: 'If for-profit organisations are to become active in social tourism, then there should be subsidies – for example an employment tax that is allocated to social tourism initiatives.'
>
> Respondent 2: 'I completely disagree – the average person pays more than enough taxes already. Public–private partnerships are what is needed to make this happen, without raising taxes by even a cent.'

While both perspectives were represented, the majority of respondents felt that private-sector involvement in social tourism would be positive, even if they considered that some public or not-for-profit provision would still be necessary to reach those with the most extensive needs. Frustration was expressed that organizations such as labour unions and mutual companies, so important in social tourism's early days, now lacked innovation. As discussed in the section about the statement activity, there was widespread scepticism about the effectiveness of these organizations in reducing travel barriers. The fact that these organizations are eligible for subsidies was an added gripe. Therefore the majority of the respondents felt that private-sector organizations with a strong social mission might provide a stimulus to social tourism, as partners in public–private partnerships.

The characteristics respondents placed at the bottom of the list were:

- inclusive hiring strategies;
- linkages with local communities;
- environmental sustainability.

The respondents felt these characteristics ideally should apply to all tourism businesses, but were not specific to social tourism alone.

12.6 Conclusions

This chapter has highlighted the complexity of defining social tourism, and stakeholders in social tourism, as illustrated by findings from a series of focus groups conducted in Flanders, Belgium. Beneficiaries of social tourism vary widely, and therefore their travel needs and wants may vary equally. Within social tourism, there may be differences between the roles the voluntary/public and private sectors play in reaching different audiences. Private-sector partners often express interest in reaching 'easier-to-serve' beneficiaries, often on a 'break-even' basis for those customers. Groups with very distinct travel needs, however, (e.g. travellers with severe disabilities) tend to be served by public/voluntary providers, in a model where subsidies or donations allow the holiday to be offered at low-cost rates.

In all focus groups, the frustration was expressed (at least by some of the respondents) that there are tourism providers in Flanders who are historically associated with social tourism, and who continue to receive government subsidies because of this reason, yet who are no longer driving innovation and practice in the social tourism field. A conclusion of the focus groups was therefore that a definition of social tourism based on the nature of the provider is not helpful: in practice, actors in the public, private and voluntary sectors are social tourism providers. A high level of consensus was reached for the statement: 'social tourism is something you DO, not something you ARE'. It should be seen as a mechanism that facilitates access to tourism and leisure to target groups who are usually excluded from them.

If social tourism is seen as a mechanism, and not a distinct subsector of tourism with its own providers, then the key goal of the mechanism is to remove barriers to travel participation. The financial barrier is often important, yet the focus groups highlighted that removing this barrier is not always enough: low-cost carriers for example may remove it to some extent, but do not usually reach the target groups served by social tourism (often because these groups lack the travel experience to find and manage a holiday without support). Social tourism thus removes at least two barriers simultaneously, by offering access to information, support, guidance and expertise.

References

Bennett, R. (1998) Corporate philanthropy in France, Germany and the UK: international comparisons of commercial orientation towards company giving in European nations. *International Marketing Review* 15(6), 458–475.

Chauvin, J. (2002) *Le tourisme social et associatif en France. [Social and Associative Tourism in France]*. l'Harmattan, Paris.

Crane, A., Matten, D. and Spence, L. (2008) *Corporate Social Responsibility: Readings and Cases in a Global Context*. Routledge, London.

Halme, M. and Laurila, J. (2009) Philanthropy, integration or innovation? Exploring the financial and societal outcomes of different types of corporate responsibility. *Journal of Business Ethics* 84(3), 325–339. DOI: 10.1007/s10551-008-9712-5.

Hess, D., Rogovsky, N. and Dunfee, T. (2008) The next wave of corporate community involvement: corporate social initiatives. In: Crane, A., Matten, D. and Spence, L. (eds) *Corporate Social Responsibility: Readings and Cases in a Global Context*. Routledge, London.

Lanquar, R. and Raynouard, Y. (1986) *Le Tourisme Social*. Presses Universitaires de France, Paris.

Maignan, I. and Ralston, D.A. (2002) Corporate social responsibility in Europe and the US.: Insights from businesses' self-representations. *Journal of International Business Studies* 33(3), 497–514. DOI: 10.1057/palgrave.jibs.8491028.

McCabe, S., Minnaert, L. and Diekmann, A. (2011) *Social Tourism in Europe: Theory and Practice*. Channel View Publications, Bristol, UK.

McWilliams, A., Siegel, D. and Wright, P. (2005) *Corporate social responsibility: Strategic implications. Rensselaer Working Papers in Economics, Number 0506*. Rensselaer Polytechnic Institute, New York.

Minnaert, L. (2011) Defining social tourism in its historical context. In: McCabe, S., Minnaert, L. and Diekmann, A. (eds) *Social Tourism in Europe: Theory and Practice*. Vol. 52. Channel View Publications, Bristol, UK.

Porter, M. and Kramer, M. (2008) The competitive advantage of corporate philanthropy. In: Crane, A., Matten, D. and Spence, L. (eds) *Corporate Social Responsibility: Readings and Cases in a Global Context*. Routledge, London.

Tourism Flanders (2018) Strategische Transities voor het Sociaal Toerisme. [Strategic Transitions for Social Tourism.] Toerisme Vlaanderen. Brussels. Available at: www.vakantieparticipatie.be/content/numbers (accessed 20 October 2020).

13 Social Tourism Policies: Critical Reflections

Anya Diekmann[1]* and Scott McCabe[2,3]
[1]Université Libre de Bruxelles, Brussels, Belgium; [2]Nottingham University Business School, University of Nottingham, UK; [3]University of Eastern Finland, Joensuu, Finland

13.1 Introduction

In 2017, an average of 30.5%[1] of Europeans did not take a holiday (Eurostat, 2018). The figures, however, reveal significant differences between countries. In Romania for example, 73.2% did not participate, in contrast to only 8.7% of Finns (Eurostat, 2018). Naturally, different cultural practices and social norms of holiday-making are likely to influence these figures, in conjunction with other factors. Yet they indicate the likelihood that social inequalities that have existed between Europeans in their ability to access opportunities for leisure travel over many decades persist despite policies and systems at the national and European Union (EU) level to support holiday-taking. While overall non-participation rates have decreased in recent years, with an increase in the overall proportion of people taking a vacation, we can observe a widening of income differences leading to rising social inequalities throughout Europe (Keeley, 2015).

Having established the foundations of the rights to tourism and travel, based in the UN's Declaration of Human Rights, most European countries are aware of the role tourism plays in providing access to global resources. The United Nations World Trade Organization's (UNWTO) article on Global Code of Ethics for Tourism seven reaffirms the basis for these rights and expressly that social tourism should be established to facilitate them for all in society:

> The prospect of direct and personal access to the discovery and enjoyment of the planet's resources constitutes a right equally open to all the world's inhabitants; the increasingly extensive participation in national and international tourism should be regarded as one of the best possible expressions of the sustained growth of free time, and obstacles should not be placed in its way.
>
> The universal right to tourism must be regarded as the corollary of the right to rest and leisure, including reasonable limitation of working hours and periodic holidays with pay, guaranteed by Article 24 of the Universal Declaration of Human Rights and Article 7.d of the International Covenant on Economic, Social and Cultural Rights.
>
> Social tourism, and in particular associative tourism, which facilitates widespread access to leisure, travel and holidays, should be developed with the support of the public authorities. (UNWTO Global Code of Ethics for Tourism, UNWTO, 2020)

It is evident that despite such expressions of international policy, the implementation of such

*Corresponding author: Anya.Diekmann@ulb.be

 Social Tourism: Global Challenges and Approaches (eds J. Lima and C. Eusébio)
DOI: 10.1079/9781789241211.0014

rights remains variable. Although currently, many countries do provide social tourism schemes and policies that aim to ensure that tourism participation is accessible to parts of the population that do not have access to holidays for a variety of reasons, the participation figures highlight only partial success in these endeavours.

The various social tourism schemes that have been devised and implemented in Europe depend to a large extent on the historical context of each country, specifically, the level of importance attached to leisure travel throughout the industrialization processes undertaken in the 20th century. The place occupied by tourism in social norms, as well as public policies for welfare, determined the public or private organizations and the range of intermediaries involved in holiday provision for the population (Diekmann *et al.*, 2012, p. 37). However, most of these can trace their roots even further back to the first half of the 19th century when workers initially gained the right to holidays from work with pay. Therefore, a critical issue is whether the types of policies for social tourism have been effective in opening up access to tourism opportunities, and the extent to which they have reduced social inequalities in participation in tourism in Europe.

A further issue is at the macro level, a range of societal, political and technological developments in recent decades, such as globalization, increased mobilities, the restructuring of economies to services from manufacturing and a reduction in costs of travel owing to technology advancement and competition; all these have opened up a vast new range of opportunities for tourism and leisure. Tourism and the associated leisure economy is one of the most important and highest-growth sectors of the economy, accounting for 10.4% of global GDP, and 9.7% in Europe specifically, producing growth of 3.1% in Europe in 2018, outperforming general economic growth and many other sectors (World Travel and Tourism Council, 2019). This proliferation and growth led to an assumption that holidays and leisure are easily available to all in society. However, this has not yielded a corresponding democratization of tourism, and these structural changes have also led to increasing inequalities and disparities in holiday participation (Cole and Morgan, 2010).

This has resulted in new challenges for the broader concept of social tourism. These transitional social and economic factors place greater emphasis on the importance of leisure and holidays, which have become further embedded as vital pillars of the European economic system. Yet mobility for disadvantaged groups is reducing, some sections of society are not able to access technological affordances and the effects of economic restructuring coupled with the aftermath of the global financial crisis have exacerbated challenges, due to pressure on public finances (Kakoudakis and McCabe, 2018). Decreasing subsidies combined with high competition from the commercial sector have compounded challenges to social tourism systems at the national level. Similarly, structural changes in society, including the ageing of the European population but also the changes to family composition and family life, have dramatically altered the types of people affected by social exclusion in tourism, revealing questions about the efficacy of established social tourism programmes for the future. This has meant that a focus on policies and their implementation is critical to understanding the context for, as well as developing mechanisms to overcome, current and future challenges.

In parallel to the ways that tourism policies and planning more generally have evolved since the Second World War (WWII), social tourism policies have been on the agenda of most European countries although to varying degrees, as a part of their welfare provision or policy. However, due to the inexorable creep of neoliberalism throughout the world, the origins of social tourism as embedded as an aspect of the welfare state system is slowly but surely being replaced by a focus on the potential it holds for destination development, and to produce tax revenue and savings in public social costs (McCabe and Diekmann, 2015). Additionally, a rise in more individualistic notions concerning the rights to holidays (Minnaert *et al.*, 2006; Minnaert, 2007), which position holidaymaking as a reward rather than an entitlement, means that social tourism has to reinvent itself to be more inclusive rather than focusing largely on blue colour workers.

Concurrently, there are many opportunities for the development of social tourism in the future, such as the increasing consideration given by national governments to the importance of 'well-being' of all citizens in modern societies. Another possibility is the increasing emphasis being placed on issues of sustainability by the

European media, raising awareness of environmental and social issues, which dovetail neatly into the agenda of policy-makers concerning the ethical and responsible development of tourism. Media attention to the plight of the planet in the face of climate change and the threat to biodiversity, made popular through TV programmes such as the BBC's *Blue Planet II* (Hayns-Worthington, 2018), and through the rise of teenage environmental activists such as Greta Thunberg (Watts, 2019) has catapulted sustainability issues up the European political agenda.

At the same time, debates about the effects of peak demand in the form of 'overtourism' (Jacobsen *et al.*, 2019) may well lead into a consideration of social tourism's role in evening out the spread of demand in destinations (Cisneros-Martínez *et al.*, 2018). The possible benefits for the tourism industry could be a refocusing on more sustainable and responsible forms of tourism in the future, of which social tourism provides a proven approach. As governments integrate well-being in their policy agendas and try to address concerns of overtourism and its effects on local population attitudes, the recognition of social tourism activities to deliver potential benefits to local populations (either as consumers or producers of social tourism services) will be a great opportunity. In that context, some governments are currently reviewing their involvement in social tourism, perhaps as a response to a growing demand among the population (Diekmann *et al.*, 2018).

This chapter examines these recent developments by analysing the factors underpinning them from a historical perspective and by contrasting the expansion in social tourism policies alongside general tourism policies. It goes on to provide a critical analysis of the various types of policies, from the system of delivery of subsidies and support for the intermediaries providing social tourism programmes to the system, which favours aid given directly to the person, and how they are implemented throughout Europe.

13.2 A Short History of the Development of Holidays for Workers

In order to understand the context driving critical issues in social tourism policy today, it is useful to briefly sketch out the background to social tourism movements from a policy and welfare perspective. Contrary to common beliefs, workers have benefited from 'holy days' and later 'blue Mondays' from the very early times of the industrialization process before the mid-19th century (Hachtmann, 2007; Zuelow, 2015) in recognition of the consequences of the Industrial Revolution, which led to a significant increase in the numbers of working days and hours (Voth, 2000). These holiday entitlements were complemented by wakes weeks and town holidays in the 19th century, engendering a social characteristic to early popular tourism (Walton, 2013). Although only a minority of workers could benefit from some of the short breaks in work, and although some still exist today, most had disappeared by the end of the 19th century (Hachtmann, 2007). Even so, they laid the basis for future social tourism development and are significant since they contradicted the idea that tourism has trickled down from the upper classes to the lower classes through the better-known examples of the 'grand tour' practised by wealthy elites, and immortalized in the arts, literature and high culture of the 18th and 19th centuries (Hachtmann, 2007). In the early 19th century, no legal framework was in place that guaranteed workers a break from work, with the exception of the Bank Holidays Act of 1871, which addressed mostly bank workers (Walton, 2013). Yet trade unions, particularly after the First World War, negotiated collective bargaining agreements and gradually, paid holidays of around 3 days each year were established. From 1919[2] onwards, the legal basis for paid holidays was introduced in most European countries, with a majority giving that right in the 1930s.

Organized tourism developed in the early 1920s through the auspices of workers' (or trades) unions. Often these were overtly political in nature. In the UK for example, the 'Workers Travel Association' (WTA) developed associative tourism opportunities aimed at allowing workers to meet workers from other countries, developing international friendship to prevent conflicts like the First World War from ever happening again.

> There were no features of pleasure for pleasure's own sake; quite to the contrary, trips were

planned to the battlefields of the Great War, the idea being to demonstrate the destruction brought by war. Other trips should be arranged in order to study the International Labour Organisation (ILO) and the League of Nations in Switzerland, and social reforms in Denmark. Holiday goers with more conventional tastes were pointed towards more ordinary tour operators. (Gråbacke, 2006, April, p. 4)

During the economic crisis between the two world wars, new policies at governmental level towards domestic tourism, with the development of holiday camps to accommodate workers' families for payable holidays, were developed and instrumentalized by different regimes, such as fascists and communists. For instance, the fascist ideology developed tourism as a tool to promote the nation and its greatness as well as the healthfulness of its people (Zuelow, 2015) by organizing trips for workers with organizations such as *Dopolavoro*[3] (Italy) and *Kraft durch Freude*[4] (Germany). In democratic countries, tourism was considered by governments to be a private matter, but workers' organizations and trade unions provided accessible holidays. For instance, in Scandinavia, this was done by RESO (Sweden), *Dansk Folke-Ferie* (Denmark) and the People's Travel Agency *Kansan Matkatoimisto* (Gråbacke, 2006, April), and in France, the trade union CGT created the *Bureau de tourisme populaire* in 1937, merging in 1398 with the travel organization *vacances pour tous*[5] created by teachers (Hachtmann, 2007, p. 109).

From mainly trade unions and private initiative (with the exception of the fascists' and communists' regimes between the two wars), after the Second World War governments were urged to regulate a growing tourism market though policies.

13.3 The Proliferation of (Social) Tourism Policies After the Second World War

After the interruption of activities due to the Second World War, tourism development rapidly increased owing in part to the expansion of jet air travel and liberalized development and international trade policies. As tourism activities took off on a broad level, policies were needed to regulate tourism and protect consumers.

However, before looking into the details, it is important to define what we mean by policies. The most frequently quoted definition of policies seems to be that of Dye (1984): 'whatever governments choose to do or not to do' (Dredge and Jenkins, 2007; Hall, 2008; Airey, 2015). While being criticized for being too simplistic, 'it captures the essence of public policy as being the outcomes of decisions or choices made, or sometimes more importantly not made, by public policy makers' (Airey, 2015, p. 247). What is relevant here is that policies translate the intentions of planning and recognizing the rights to holiday participation in welfare states. Yet the way these policies are implemented depends on all sorts of external factors. These external factors include, among others, economic development, labour markets, political orientations and welfare systems, all intertwined to set the backdrop of social tourism policies that developed accordingly. These factors, which are inherently different to each country, led to a myriad of policies depending on a variety of stakeholders in different countries. However, at a supranational level, the European Commission published policies that aim at regulating access to holiday participation at the European level with their own implementation programme (see below). For instance, public social tourism policies can be at different levels, such as European, national, federal, regional or local, either only at one level (e.g. Belgium, where social tourism policies are only at the regional level), or, as is the case, for instance, in Spain at various levels, national and regional. In addition, the responsible agencies/stakeholders are not the same throughout Europe. While in some countries, the Ministry of Tourism (regional or national) is in charge, in others it may be the Ministry of Social Affairs, etc. Policies are then implemented in various ways, either directly or through programmes and/or intermediaries (Diekmann *et al.*, 2012), as will be discussed later.

Table 13.1 provides an overview of how the development and systematic integration of policies on tourism can be compared with the development of policies and initiatives on social tourism, independent of the level (EU, national, or regional). In some respects, these policy developments mirror or reflect changing external environment conditions or trends.

The post-war period witnessed increasing industrial development worldwide that led to

Table 13.1. Tourism and social tourism policies. (From Authors and Hall, 2008, p. 45).

Phase	Characteristics Tourism (this column is entirely cited from Hall, 2008, p. 45)	Social tourism
1945–1955	Dismantling and streamlining of the policy, customs currency and health regulations that had been put into place following the Second World War	Governments subsidize social tourism infrastructures depending on trade unions, workers' councils, health care organizations and popular educational movements and collectives, (Chauvin, 2002) building subsidies
1955–1970	Greater government involvement in tourism marketing in order to increase tourism earning potential	Several governments integrate 'social tourism' into their social welfare policies, resulting in state-backed schemes to provide and promote affordable holidays and recreation through various distribution channels, mostly through intermediaries (Diekmann *et al.*, 2012)
1970–1985	Government involvement in the supply of tourism infrastructure and in the use of tourism as a tool of regional development	Several national programmes addressing workers are launched depending on public authorities, such as IMSERSO in Spain (see below)
1985–2000	Continued use of tourism as a tool for regional development, increased focus on environmental issues, reduced direct government involvement in the supply of tourism infrastructure, greater emphasis on the development of public–private partnerships and industry self-regulation	Focus on social cohesion and inclusion in social tourism policies and focused policies on 'new' target groups
2000–2010	Continued use of tourism as a tool for regional development; greater focus on network development, collaboration and clustering; security and crisis management new dimensions of tourism policy. Environmental issues such as climate change are prominent along with broader issues of global environmental change	Sustainability becomes an intrinsic part of social tourism policies. An example is the integration of social tourism issues in the policy remit of the Tourism Sustainability Group (TSG) Definition of the European Economic and Social Commission of tourism as a right in the 'Renewed Tourism Policy' (COM, 2006) European Union (EU) puts social tourism on the agenda through its preparatory action Calypso (2008) with an economic focus on lower seasonality consolidating the four target groups
2010–present	Consumer policies at EU level governing the rights and obligations of the users of tourism services Enforcing of sustainable and responsible tourism policies Regulations of new tourism supply models such as Airbnb New policies to fight 'overtourism'	Tourism for All policies at EU level, yet depending on national governments Creation of the Commission on Social Policies for Tourism

an increase in activity and influence of trade unions and health care organizations in welfare states and their involvement in holiday provision in many European countries, whereby unions bought property used by their members, and effectively became tourism businesses through ownership of accommodation or by building holiday villages and camps.

In the 1960s and 1970s, the competitive commercial tourism supply increased with the development of mass tourism destinations and the famous sea, sand and sun, allowing workers to benefit equally from other offers in other countries. The consequences were a decrease in and often reorientation of social tourism providers towards a commercial market, however integrating sustainability issues.

However, social tourism policies generally sought to counterbalance the mass development of market-driven tourism activities, and social tourism policies prioritized health and social inclusion against customs and security issues (see Table 13.1). From the original focus on workers, a shift in policies occurred to integrate specific other target groups (youth, families, the disabled and seniors) consolidated by the EU's preparatory action of Calypso in 2008. Social tourism policies thus focus in general on structuring access to holidays and programmes for specific target groups or on the public funding for social tourism provision. In opposition to social tourism policies, general tourism policies only exceptionally consider tourists. An example would be the recent consumer policies at the EU level that govern the rights and obligations of the users of tourism services (European Commission (EC), n.d).

When sustainability came onto the agenda, policy-makers at the European level made sustainability a core value of social tourism. It should, however, be underlined that none of the general or social tourism policies specifically considered the host community until recently, when social tourism policies at the international level incorporated a fair tourism approach to protecting host communities (ISTO, n.d).

While there are a great variety of social tourism policies, some common (or at least comparable) characteristics, such as the funding or type of subsidies can be observed as highlighted in the following section.

13.4 Different Types of Subsidies

Before looking into the different types of subsidies, it is important to note that all social tourism funding schemes relate to domestic tourism activities (only a few European funded programmes allow funded travel to another country). Most of the countries that have implemented a governmental social tourism scheme originally developed policies that are either building subsidies or person subsidies. Increasingly, owing to a shift in target groups, both schemes are used jointly. The first type of subsidy is distributed by governments to different types of organizations fulfilling some legal requirements for hosting and accommodating specific target groups. Most of the recognized social tourism organizations are either private or (semi-)public non-profit associations providing accommodation facilities, such as holiday centres, hotels, youth hostels, camping sites etc. In the Belgian region Wallonia, for example, only buildings belonging to recognized social tourism providers are eligible to receive money from the Walloon government. These providers are connected to the trade unions and health care organizations. In the tourism policies launched in 2010 (Code Wallon du tourism, 2010) the type of beneficiaries is stipulated in Art. 12:

> Leisure activities and holidays organised by an association in order to offer any person, and in particular to people with economic or cultural disadvantages, best practical conditions to allow real access to the activities.[6]

However, in most countries, social tourism accommodation, such as holiday villages, hotels, youth hostels and camping sites are open to everybody, but legal requirements include, for example, the allocation of a majority of rooms (e.g. 51% in Wallonia) to members of the organization, generally blue-collar workers and employees.

Reasons for the building subsidies approach are historical as, after the introduction of paid holidays, trade unions and health care organizations were among the first to provide holiday accommodations to workers without any previous holiday experience (Walton, 2013). It is consequently not surprising to note that the main user group benefiting from the subsidies is the working middle class, which is of course connected to the ownership of the facilities, either to workers' unions or to public health care organizations (Caire, 2007). In addition, the competition of the commercial sector and the maintenance costs of the buildings, often built in the late 1970s or 1980s, leads to uncompetitive prices and excludes disadvantaged people, so missing out the core value of social tourism: access for all (Diekmann and Bauthier, 2011).

From the 1980s onward, a new system giving more freedom of choice of destination (although still domestic) to workers saw the light: the person subsidy. This new approach of individual aid for holiday participation developed to its full extent with a shift from providing holidays to workers to a more inclusive approach targeting different groups.

The person subsidy often translates into either:

1. Funding going directly to the beneficiary (direct person subsidy).
2. Funding going to a public or private organization offering holidays at reduced prices (indirect person subsidy).

Direct Person Subsidies (DPS) include, for instance, holiday vouchers, such as operated in France since 1982 by a state agency ANCV (*Agence Nationale pour les Chèques-Vacances*), where every employee receives a voucher exchangeable for various leisure activities, but mainly aimed at paying (at least partially) for holidays. This scheme related to the historical approach towards social tourism support, which supported people in employment, but excluded disadvantaged unemployed people. In order to lower the so-called 'tourism divide', the agency created a solidarity scheme to help disadvantaged and fragile persons to go on holiday, especially related to the four target groups: families (with a particular attention to mono-parental families), seniors, people with disabilities and young people. Specific programmes have been created to support target groups. For example, since 2002, *Sac Ados*[7] has used vouchers to fund travel for young people generally excluded from holidays (ISTO, 2017). It should be stressed that ANCV is also owner of 14 holiday centres benefiting from building subsidies from the French national government. In that perspective, it is a combination of BS and DPS managed by the same authority. A clear shift of the subsidies from BS towards DPS can be observed from the 1980s on (Caire, 2011).

This combination of two types of subsidies can also be observed in the Belgian region of Flanders. In addition to the long existing BS (as in Wallonia), since 2003, the decree 'Tourism for All' and the establishment of government services such as the Holiday Participation Centre has been set up to stimulate holiday participation among people and families with low income (Schapmans, 2012). In their Information Point for Accessible Tourism, the Flemish government provides overall information and labels accessible accommodation, putting more emphasis on the participation of target groups (especially people living in poverty and people with disabilities (Gijselinckx and De Smet, 2013, p. 43).

The increasing concern with poverty has also led to a similar system on the Walloon side in addition to the existing BS. In 2018, the government launched a DPS to allow holiday and leisure participation for disadvantaged people within the framework of a broader policy on combatting poverty (Wallonie, n.d). Through income-related reduced prices, individuals can thus choose to participate in partnering leisure activities and holiday provision.

In Spain and Portugal, public agencies depending on ministries implement their social tourism policies through specific target group programmes and indirect person subsidies (IPS). For instance, since 1985 in Spain, the government has funded applicants based on their income. The funding goes to IMSERSO, which redistributes it indirectly through a variable price reduction to selected beneficiaries who then can choose to enjoy a break in any commercial tourism accommodation. The system in Portugal is very similar. Both programmes, promoted by IMSERSO in Spain and INATEL in Portugal, were created in the mid-1980s with the aim of allowing seniors to benefit from more accessible holidays in the lower season (benefiting the tourism sector of destinations). Another option is the combination of IPS and DPS. Both coexist in the Hungarian *Erzsebet programm* where, with the help of vouchers (DPS), among other target groups, disadvantaged children and families can participate in a camp holiday organized by the *Erzsebet foundation* funded by the government (BS), contributing to social inclusion (ISTO, 2017). This programme was launched in 2012 and can be considered representative of an innovative approach combining both funding schemes.

In summary, almost all countries starting early with social tourism policies developed a building subsidy approach as the governments relied on the workers' unions and health care organizations for the provision of holiday participation and accommodation for their members. The first PS were introduced for workers in the 1980s with different sets of implementation schemes, either direct through vouchers or indirect through price reductions. With increasing competition from the commercial sector offering inexpensive holidays in other destinations and an increasing exclusion of certain groups of the population, new target groups emerged, leading to the need to review existing funding schemes. In that perspective, new policies were launched focusing on people excluded from holiday participation. The PS have significantly increased over the past two decades, often as an addition to existing BS schemes. The financial crisis in 2008 can be considered as one of the triggers for the reconsideration of social tourism policies acknowledging the right to tourism. For example, although Spain suffered greatly from the crisis and the country had to reduce public expenses, the public funding of the IMSERSO programme continued without interruption.

13.5 Conclusions and Future Perspectives

Despite a general perception that in recent years, or at least in the current decade, there has been a decreasing governmental support for social tourism in Europe, specifically in the context of the global financial crisis, the present chapter shows the reverse to be true. Indeed, policies and programmes designed to support holiday participation as part of social inclusion measures not only continue to exist but have even increased. At the European level, for instance, over the past 10 years, increased attention has been given to developing systems of support to allow access to holidays for all. What has changed has been a reshaping of social tourism policies from a system whereby support is provided to accommodation or facilities to one where the main support is directed to the person, the social tourist. Thereby, some aspects of the support have been reduced, but have been replaced by other financial mechanisms. However, one problematic aspect of this shift is that traditional social tourism intermediaries, such as the unions and health care organizations, must reconsider their contribution to social tourism and adapt their offer more to specific target groups, thus integrating the PS and BS.

Innovative participatory schemes are developed in many countries and mutual inspiration and collaboration between the social tourism stakeholders should consolidate social tourism policies at all levels. What is therefore needed in the near future is a tool to enable the analysis of social tourism policies, such as an observatory allowing an understanding of the different modes of delivery and systems of support for social tourism, including a deeper insight into subsidy mechanisms.

An encouraging sign of the increasing significance of collaborative approaches and need for social tourism policies is the freshly created Commission on Social Policies for Tourism (April 2019), which aims at creating a place of debates, exchanges and proposals in order to allow public authorities and other concerned organizations to know existing mechanisms better, to measure their strengths and weaknesses and to promote them to those wishing to acquire such devices to facilitate access to 'Tourism for All' and the development of responsible tourism. Such a pan-European approach will help to strengthen policies.

Another challenge for tourism, but social tourism in particular, is how to deal with overtourism or increasing pressure on destinations. Social tourism stakeholder value sustainability is one of the core features of social tourism practices. So how can participation be increased and helped in a sustainable way?

Notes

[1] Aged 15 or over of the European Union, with at least 1 overnight stay.
[2] Austria was the first country to introduce paid holidays, followed by Poland, Czechoslovakia, USSR and Luxembourg.
[3] Translation: after work
[4] Translation: Strength through joy
[5] It is interesting to mention that the term 'holidays for all' is again broadly used today.
[6] Art. 313. D provides a definition of 'association providing as social tourism accommodation'. The conditions for subsidies for the social tourism accommodations are described in art. 414.
[7] A word game relating to backpacking and youth, could be translated as Teenage Backpack

References

Airey, D. (2015) Developments in understanding tourism policy. *Tourism Review* 70(4), 246–258. DOI: 10.1108/TR-08-2014-0052.

Caire, G. (2007) Tourisme solidaire, capacités et développement socialement durable. [Solidarity tourism, capacities and socially sustainable development]. *Marché et organisations* 1, 89–115.

Caire, G. (2011) Social tourism and the social economy. In: McCabe, S., Minnaert, L. and Diekmann, A. (eds) *Social Tourism in Europe: Theory and Practice*. Channel View Publications, Bristol, UK, pp. 52–73.

Chauvin, J. (2002) *Le tourisme social et associatif en France: Acteur majeur de l'économie sociale. [Social and Associative Tourism in France: A Major Player in the Social Economy]*. Editions L'Harmattan, Paris.

Cisneros-Martínez, J.D., McCabe, S. and Fernández-Morales, A. (2018) The contribution of social tourism to sustainable tourism: a case study of seasonally adjusted programmes in Spain. *Journal of Sustainable Tourism* 26(1), 85–107. DOI: 10.1080/09669582.2017.1319844.

Code Wallon du tourism (2010) Homepage. Available at: https://wallex.wallonie.be/contents/acts/19/19701/1.html?doc=17080&rev=17977-4010 (accessed 26 October 2020).

COM (2006) *A Renewed EU Tourism Policy: Towards a stronger partnership for European Tourism*, final of 17 March 2006. COM, p. 134.

Diekmann, A. and Bauthier, I. (2011) *Tourisme Social en Wallonie. Réalités et Perspectives. [Social tourism in Wallon. Realities and perspectives.]*. Rapport pour le CGT, CGT.

Diekmann, A., McCabe, S. and Minnaert, L. (2012) Social tourism today: stakeholders, and supply and demand factors. In: McCabe, S., Minnaert, L. and Diekmann, A. (eds) *Social Tourism in Europe: Theory and Practice*. Channel View Publications, Bristol, UK, pp. 35–47.

Diekmann, A., McCabe, S. and Ferreira, C.C. (2018) Social tourism: research advances, but stasis in policy. bridging the divide. *Journal of Policy Research in Tourism, Leisure and Events* 10(3), 181–188. DOI: 10.1080/19407963.2018.1490859.

Dredge, D. and Jenkins, J.M. (2007) *Tourism Planning and Policy*. John Wiley & Sons Australia, Milton, Australia.

Dye, T.R. (1984) Party and policy in the states. *The Journal of Politics* 46(4), 1097–1116. DOI: 10.2307/2131244.

European Commission (EC) (n.d) The regulation of tourism activity in Europe. Available at: https://ec.europa.eu/growth/sectors/tourism/business-portal/understanding-legislation/regulation-tourism-activity-europe_en#specific (accessed 20 October 2020).

Eurostat (2018) Tourism statistics. Available at: https://ec.europa.eu/eurostat/statistics-explained/index.php/Tourism_statistics (accessed 20 October 2020).

Gijselinckx, C. and De Smet, L. (2013) Holidays with a plus. The multi-faceted reality of social tourism in Flanders unveiled. In: Diekmann, A. and Jolin, L. (eds) *Regards croisés sur le tourisme social dans le monde: L'apport de la recherche*. Presses de l'Université du Québec, Québec, Canada.

Gråbacke, C. (2006, April) Organised leisure for the working class – European popular travel and leisure organisations in the interwar period. *Presented at Economic History Society Annual Conference*, 31 March – 2 April 2006.

Hachtmann, R. (2007) *Tourismus-Geschichte*. Vol. 2866. UTB, Göttingen, Germany.
Hall, C.M. (2008) *Tourism Planning: Policies, Processes and Relationships*. Pearson Education, London.
Hayns-Worthington, S. (2018) The Attenborough effect: searches for plastic recycling rocket after blue planet II. Available at: https://resource.co/article/attenborough-effect-searches-plastic-recycling-rocket-after-blue-planet-ii-12334 (accessed 20 October 2020).
ISTO (2017) Tourism in Actions – 20 examples of social policies & programmes around the world. Available at: www.oits-isto.org/oits/public/section.jsf?id=39 (accessed 20 October 2020).
ISTO (n.d) Homepage. Available at: www.oits-isto.org (accessed 20 October 2020).
Jacobsen, J.K.S., Iversen, N.M. and Hem, L.E. (2019) Hotspot crowding and over-tourism: antecedents of destination attractiveness. *Annals of Tourism Research* 76, 53–66. DOI: 10.1016/j.annals.2019.02.011.
Kakoudakis, K.I. and McCabe, S. (2018) Social tourism as a modest, yet sustainable, development strategy: policy recommendations for Greece. *Journal of Policy Research in Tourism, Leisure and Events* 10(3), 189–203. DOI: 10.1080/19407963.2018.1443938.
Keeley, B. (2015) *Income Inequality: The Gap between Rich and Poor, OECD Insights*. OECD Publishing, Paris.
McCabe, S. and Diekmann, A. (2015) The rights to tourism: reflections on social tourism and human rights. *Tourism Recreation Research* 40(2), 194–204. DOI: 10.1080/02508281.2015.1049022.
Minnaert, L. (2007) Social tourism: a potential policy to reduce social exclusion? The effects of visitor-related social tourism for low-income groups on personal and family development. Doctoral dissertation. University of Westminster, London.
Minnaert, L., Maitland, R. and Miller, G. (2006) Social tourism and its ethical foundations. *Tourism Culture & Communication* 7(1), 7–17. DOI: 10.3727/109830406778493533.
Schapmans, M. (2012) *Holiday Participation Center – Because everyone needs a break*. Presentation at the Netstar Seminar, UK.
UNWTO (2020) Global code of ethics for tourism. Available at: https://www.unwto.org/global-code-of-ethics-for-tourism (accessed 26 October 2020).
Voth, H.J. (2000) *Time and Work in England 1750–1830*. Clarendon Press, Oxford, UK.
Wallonie (n.d) Homepage. Available at: https://www.wallonie.be/fr/actualites/un-nouveau-plan-wallon-de-lutte-contre-la-pauvrete (accessed 20 October 2020).
Walton, J.K. (2013) 'Social tourism' in Britain: history and prospects. *Journal of Policy Research in Tourism, Leisure and Events* 5(1), 46–61. DOI: 10.1080/19407963.2012.703377.
Watts, J. (2019) The Greta Thunberg effect: at last, MPS focus on climate change. Available at: https://www.theguardian.com/environment/2019/apr/23/greta-thunberg (accessed 20 October 2020).
World Travel and Tourism Council (2019) Travel and tourism economic impact 2019 report. Available at: https://www.wttc.org/-/media/files/reports/economic-impact-research/regions-2019/world2019.pdf (accessed 20 October 2020).
Zuelow, E. (2015) *A History of Modern Tourism*. Macmillan International Higher Education, London.

Social Tourism – Global Challenges and Approaches in the Future

Celeste Eusébio[1]* and Joana Lima[2]
[1]Department of Economics, Management, Industrial Engineering and Tourism, University of Aveiro and GOVCOPP Research Unit, Aveiro, Portugal; [2]Social Sciences School, University of Évora and CIDEHUS Research Unit, Évora, Portugal

General Overview and Main Contributions

This book analysed social tourism across the world, exploring the theoretical foundations of this type of tourism with examples of practical implementations. Social tourism has a long history and is a dynamic, complex and multifaceted concept, and therefore both conceptual chapters and chapters with practical examples were included in this book. The contents of this book permit us both to look back and to identify strategies to move forward.

Although the relevance of social tourism is widely recognized worldwide owing to its important benefits for people, economies and society, considerable differences concerning the offer of social initiatives can be observed across the world. Different countries offer different social initiatives. Moreover, while this type of tourism is well established in some parts of the world, it is almost unknown in others.

When people have the chance to carry out tourism trips, they have the opportunity to escape from routine, to rest, and to obtain new experiences and knowledge. Consequently, participation in tourism activities has the potential to promote development in the personal, health and social spheres. However, a great number of people have not the chance to participate in tourism activities, given the various travel constraints that they face. Social tourism has a great potential to promote social inclusion; it creates stronger families and improves well-being. However, despite the increase in interest observed in recent years from academics as well as tourism industry and governments, social tourism has received little attention in the tourism literature and in government policies in some parts of the world, and indeed has not received the attention it deserves. Consequently, despite an increased number of academic studies about this type of tourism in recent years, little is still known about how social tourism can improve the lives of disadvantaged groups (seniors, low-income families, young people and people with disabilities) and concerning its economic, environmental and social impacts. In this book, these four disadvantaged groups were examined in order to identify their needs, motivations and travel constraints, as well as the characteristics of some social tourism initiatives/programmes that exist in various countries for these target groups.

Theoretical and practical contributions of the content of this book may be identified. The multidisciplinary perspective adopted, the various approaches and methodologies used, and the combination of theory and practical examples across the world and in terms of target groups of

*Corresponding author: celeste.eusebio@ua.pt

DOI: 10.1079/9781789241211.0015

social tourism initiatives are identified as innovative aspects of the book. This book thus provides:

- Knowledge about the different approaches used around the world to analyse this kind of tourism.
- Insights concerning the connection between scientific knowledge and practical examples of social tourism programmes and initiatives offered across the world.
- An overview of various practical examples of programmes/initiatives across the world for the four main target groups of social tourism (low-income families, young people, senior citizens and people with disabilities).
- Insights regarding the most important challenges that this kind of tourism faces nowadays.
- Knowledge about different funding mechanisms and ideas on how this type of tourism programmes can be financed and designed in the future.

Retrospective Overview – A Summary of the Most Important Contributions of Each Chapter

This overview of the most important challenges and approaches analysed and discussed in each chapter of this book highlights differences in terms of country and social tourism's target groups. Moreover, in various chapters the changes that have occurred in recent decades in terms of social tourism initiatives are mentioned. While in some countries the number of beneficiaries of some social tourism programmes has registered a decrease, mainly as a consequence of financial crises, the awareness of the need to allow everyone to participate in tourism activities, regardless of their characteristics and conditions, has increased. Another important conclusion obtained is related to the type of tourism activities included in the social tourism programmes. The majority are domestic tourism activities, with only a very limited number of social initiatives funding international trips.

The chapters included in the book were categorized into three groups, according to the theme analysed. The first group (Part II) included chapters concerning the evolution, opportunities and challenges of social tourism in three countries (Greece, Mexico and Finland), with different approaches and challenges in terms of social tourism. In Part III of the book, the experiences lived by different target groups of social tourism were analysed. The first three chapters of this part examined social tourism initiatives for seniors in three European countries (Portugal, Spain and Slovenia) that have a great tradition in terms of social tourism programmes for seniors, mainly in the case of Portugal and Spain. The remaining four chapters in this part of the book analysed low-income families, children with disabilities and children at risk of poverty. Finally, the three chapters included in Part IV of the book discussed practical challenges and policy implications for developing social tourism in the future.

A more detailed analysis of the main conclusions of each chapter included in this book provides important inputs to identify new challenges and approaches and topics for further research. Kakoudakis and Papadoulaki (Chapter 1) presented and discussed the development of social tourism in Greece over time. Despite the long tradition of social tourism in this country, which can be traced back to the interwar years, this topic has been greatly neglected in the international literature on social tourism. Greek social tourism has developed within a politically, economically and socially turbulent environment, mainly characterized by instabilities and frequent political changes. Nevertheless, there has been an increase in interest in this type of tourism in this country since the early 1980s, when it was incorporated into the country's social welfare policy. Therefore, social tourism initiatives nowadays are mainly supported by the public sector, with a significant contribution of private organizations and labour unions.

The history of social tourism in the past 45 years in Mexico was analysed in Chapter 2, highlighting the crucial relevance of this type of tourism for Mexican citizens owing to the limited financial resources they have to spend on leisure and tourism activities. Although the history of social tourism in Mexico is linked to the creation of Turissste in 1972, there has been no continuous growth of this type of tourism. For example, by the early 1980s, social tourism programmes began to lose importance in public agencies. In the mid-1980s, new plans and programmes were developed based on European examples. Consequently, social tourism in Mexico nowadays, as well as in other countries (e.g. Portugal and Spain), includes domestic tourism trips for four groups with special needs in terms of access to tourism: families, senior citizens, youth and people with disabilities. However, the future of this type of tourism will be strongly related to the availability of financial resources.

The challenges and opportunities for the development of social tourism in Finland were analysed in Chapter 3. As a part of the national social policy, social holidays in this country are intended for people who otherwise would not have the opportunity to have a holiday away from home. In this vein, the most important targets of social tourism initiatives have been families with children, individuals and families suffering from a certain health issue, illness or disability, lonely individuals, seniors and the unemployed. However, although social tourism has a long tradition in Finland, a great number of potential beneficiaries do not participate in these social initiatives owing to an absence of knowledge of its existence. Therefore, efforts should be made to make these initiatives known to the general public.

Chapter 4 opened a set of chapters presenting social tourism initiatives for different target groups in different countries. Carneiro, Eusébio, Kastenholz and Alvelos analysed the social tourism programmes offered by the INATEL Foundation to seniors and the benefits that the seniors obtained from participation in these programmes. Portugal has a long tradition in the offer of social tourism programmes for seniors through the INATEL Foundation. In this chapter, the results presented and discussed show that the social tourism programmes promoted by this institution play a very important role in increasing the participation of seniors in tourism activities, permitting the participation of those frequently excluded from access to leisure and tourism owing to financial constraints. The large set of benefits that seniors attributed to their participation in INATEL programmes show their success and justify the corresponding investments. However, the decrease in the number of participants observed in recent years highlights the need for further stimulation of these social tourism initiatives.

Social tourism programmes for seniors in Slovenia were examined in Chapter 5, in terms of characteristics and benefits of these programmes and their potential for future development. In this country, social tourism has yet to emerge as a theme in public discussion or in research. The chapter presented results from analytical and empirical qualitative research (semi-structured interviews) among the different actors dealing with social tourism for Slovenian seniors. Three main groups of benefits were identified: (i) social inclusion/socializing; (ii) widening horizons/education; and (iii) the preservation of physical and mental health, all contributing to long-term active and healthy ageing, and, consequently, often maintaining autonomy and independence in old age. The research also highlighted the importance of specific adjustments to these programmes that enable greater access for seniors, mostly in consideration of (i) (poor) socioeconomic status; (ii) health/mobility issues; and (iii) travel uncertainty. Finally, proposals for future development and systematic measures in the field of social tourism for seniors were discussed.

The Spanish experience in terms of social tourism for seniors was examined in Chapter 6. Spain is a country with a great tradition in social tourism programmes. In Spain, the Institute for the Elderly and Social Services (IMSERSO) has been working for more than 40 years to improve the quality of social services aimed at meeting the needs and demands of elderly and dependent people. Various actions and programmes related to independence, autonomy, active ageing and spa tourism have been used by IMSERSO to accomplish these aims. González, Vila and Sánchez concluded that despite the considerable relevance of the social tourism programmes for seniors offered by IMSERSO in the past few decades, new social tourism programmes for the senior market should be developed to satisfy the needs of the actual senior tourists, who currently differ considerably from the senior tourists of a couple of decades ago.

The needs, benefits and travel constraints of a sample of Portuguese low-income families with children with disabilities were analysed in Chapter 7. This group is one of the target groups of social tourism initiatives that faces most travel constraints. The results presented and discussed in this chapter reveal that, for the majority of these families, participation in tourism activities is still only a dream, given that social tourism programmes for this market are very scarce. Economic factors, the lack of accessibility of destinations and the lack of adequately trained staff in the tourism industry are the main constraints faced by these families. However, when it is possible to carry out family tourism activities, the benefits obtained are high, not only for the well-being of the family, but also for children with disabilities, helping them to face certain challenges. Therefore, it is of the utmost importance to increase the offer of social tourism initiatives to this target group.

Children with disabilities are also frequently excluded from tourism activities owing to the various travel constraints that they face as well as the scarcity of accessible tourism services. In Chapter 8, Shiraani, Carr and Rutherford discussed how to create a space for children with disabilities in tourism, by calling for their voice to be heard and included. Listening to the voice of children with disabilities concerning their needs, desires and motivations in terms of tourism activities is of great importance in designing adapted tourism services, as well as promoting the rights of these children. Although children with disabilities are active social agents with a voice, little is known about their needs and desires, given that they have been virtually excluded from tourism research. This chapter emphasized the need to increase research in the tourism field including the view of children with disabilities. Their involvement in tourism research has the potential to aid in designing inclusive tourism policies, products, services and practices that are child-friendly and relevant to this cohort of children.

Following the same line of thought, in Chapter 9, Loureiro, Eusébio and Figueiredo highlighted the absence of children's vision and voice from tourism literature and from poverty and social exclusion studies. The majority of studies in both fields regarding children have been conducted from an adult-centric, family-centric or household-centric perspective, neglecting the children's own experiences and perspectives. Although the absence of children as the main subjects of research may be partially explained by ethical and methodological difficulties, children's feelings and experiences are quite different from those of adults, and deserve to be examined and understood in greater depth. In the same vein, this absence tends to be more evident when considering social tourism programmes, which are frequently designed to promote social inclusion of economically and socially excluded and disadvantaged children. In this chapter the authors aimed to fill these gaps in the literature by adopting a child-centric perspective regarding their participation in the holiday camps promoted by the 'O Século' Foundation (created in 1927 in Portugal), simultaneously aiming at revealing children's motivations to participate, their levels of satisfaction and the perceived benefits. Through the analysis of 131 questionnaires and 132 drawings, the chapter discussed the benefits of social tourism programmes for children from their own perspectives and experiences. Results demonstrated the children's extremely positive emotions regarding their experience in the holiday camps, mainly based on the opportunities to improve social interaction skills and create relationships of friendship and the moments of fun and happiness provided. The results equally highlight the significant contribution of social tourism programmes in promoting the well-being, happiness and social inclusion of economically disadvantaged children.

The part of the book devoted to tourism experiences lived by different target groups of social tourism (Part III) ended with a discussion of the financial accessibility of children's camps in France (Chapter 10). In this chapter, Michel highlighted the decrease in the number of children participating in summer camps in recent decades. In order to understand this phenomenon, the links between social tourism and children's camps in France were explored in the first part of the chapter, assuming that the current camps must be analysed towards the models of social tourism. The second part of the chapter examined the financial issues and outlined the camps' characteristics, giving the price levels. One way to increase the number of participants in summer camps would be to reduce the camp rates. However, this would probably involve a loss in their values, objectives and attractiveness.

Chapter 11 is the first in a group of three chapters (Part IV of the book) about practical challenges and policy implications in the scope of social tourism. In this chapter, Lima and Eusébio analysed the knowledge and interest of

Portuguese low-income families in participating in social tourism programmes. The results obtained and discussed in this chapter reveal that, although the families interviewed are interested in participating in social tourism programmes, they had never yet participated in these programmes, mainly because they are unaware of the existence of these types of programmes. In fact, the offer of social tourism programmes for low-income families is very limited in Portugal. Therefore, it is very important to increase the offer of social initiatives for this target market. This chapter also provided relevant guidelines to develop social tourism programmes for low-income families. Specifically, these programmes should be: (i) financially sustainable, both for the promoter and for families; (ii) able to adapt their offer to each type of family; (iii) monitored over time; and (iv) able to adapt their marketing strategy (particularly in terms of communication) to different types of low-income families.

In Chapter 12, Minnaert examined the roles of the public, private and voluntary sector in social tourism provision in Flanders, Belgium. In this chapter, the author remarked on the need for social tourism initiatives to become less dependent on state funding and be financially self-sustaining. Moreover, the contributions of corporate social responsibility (CSR) to stimulating social tourism initiatives were also discussed and the complexity of defining social tourism was highlighted. The existence of a great number of social tourism beneficiaries with different needs, desires and travel constraints has contributed to this complexity. Consequently, the actors involved in the supply of social tourism initiatives and the roles of these actors vary according to the country and the type of beneficiary. For example, groups with very distinct travel needs, such as travellers with severe disabilities, tend to be served by public/voluntary providers, in a model where subsidies or donations allow the holiday to be offered at low-cost rates. The insights provided by this chapter are of the utmost importance in thinking of and developing new business models for solving social problems through the participation in tourism activities of target groups who are usually excluded from them.

Finally, in Chapter 13, Diekmann and McCabe provided a critical reflection concerning social tourism policies. The authors remarked on the existence of great differences in social tourism schemes in European countries. These differences may be related to the historical context of each country, the relevance of the tourism sector and the importance of tourism in social norms, as well as in public policies for welfare. The results and discussion presented in this chapter highlighted that, although the general perception is that a decrease has occurred in governmental support for social tourism in Europe in the current decade, mainly as a consequence of the global financial crisis, in reality there has been an increase in interest in developing systems to promote 'Tourism for All'. Therefore, in many countries innovative participator schemes involving different kinds of social tourism stakeholder have been developed. According to the authors, 'what is therefore needed in the near future is a tool to enable the analysis of social tourism policies, such as an observatory allowing an understanding of the different modes of delivery and systems of support for social tourism, including a deeper insight into subsidy mechanisms'.

Prospective Overview – Challenges and Approaches in the Future

According to the topics analysed in this book, social tourism faces several challenges that need to be overcome in order to increase its number of beneficiaries. As Bélanger and Jolin (2011, p. 481) mentioned, 'social tourism sector is at the core of a humanistic vision for tourism, represents millions of people and is one of the most promising niches of recent years'. Therefore, it is relevant to identify not only future challenges but also the strategies and approaches that should be adopted to overcome these challenges.

The great social, political, economic and technological changes that have taken place over the past decade and that will be intensified in the future, will have a considerable impact on the development of social tourism initiatives worldwide. Some of these changes may be threats, while others may be opportunities. The major challenge is to develop strategies to promote social tourism that, on the one hand, will allow opportunities to be created and, on the other hand, threats to be overcome. Moreover, the pandemic situation that we are currently going through as a result of COVID-19 will be one of the main challenges that social tourism will face in the near future. While on the one hand, the state budget available for these programmes will tend to decrease, owing to the financial constraints that governments will go through as a result of this crisis, on the other hand, the number of people in need of (mainly financial) support to participate in tourism activities will tend to increase. These changes have forced the social tourism sector to reposition itself.

Given the nature of the financial schemes used in a great number of countries to fund social tourism initiatives, one of the major challenges for the development of social tourism is to identify how to become less dependent on state funding. Therefore, innovative business models should be developed, favouring the involvement of various actors/stakeholders, such as the state, volunteers and social organizations that support people with special needs and enterprises, mainly belonging to the tourism sector. These models should focus not only on the diversity of funding sources, but also on the adoption of suitable strategies that contribute to programmes becoming financially self-sustaining. Moreover, more efforts should be made to include social tourism initiatives in CSR programmes, stimulating the contribution of the private sector to the development of social tourism.

Another challenge is related to the global demographic ageing process, mainly observed in developed countries, and the emergence of new generations of senior tourists. Thus, changes in the profile of seniors require differentiated offers. Additionally, the heterogeneity of the social tourism targets groups may possibly increase in the future. Consequently, more flexible forms of social tourism initiatives should be provided.

Despite all the social tourism initiatives and programmes that have been developed in various countries and for different target groups in recent decades, there are still large numbers of people excluded from having tourism experiences. Therefore, further progress in this field must be realized to promote 'Tourism for All'. In this vein, it is essential to increase knowledge in this field, not only through theoretical work but also applied research.

Although there has been a considerable increase in research on social tourism in recent years, particularly in European countries, there are still many topics that have been little explored or almost neglected. As Diekmann *et al.* (2018, p. 181) mention, 'whilst the research emerging may be characterised as nascent, exploratory or focusing on a particular aspect of social tourism, it is without doubt that there has been widespread, inter-disciplinary interest in the concept worldwide'. In spite of the advances and awareness of the benefits of tourism participation in most countries, there is still much confusion and doubt concerning the definition, role and purpose of social tourism (Diekmann *et al.*, 2018).

The research carried out on social tourism, mainly by academics, should be used to monitor existent social tourism policies and programmes as well as to develop new policies and new social tourism initiatives. Knowledge transfer between academics, government, and public and private organizations involved in social tourism initiatives is essential to their success.

In order to increase knowledge in the field of social tourism and to overcome some research gaps, some themes where it is important to intensify research are suggested below, specifically studies to:

- Explore the potential for learning and change in behaviour through participation in tourism initiatives and evaluate the outcomes of social tourism programmes in terms of experiential learning.
- Identify the factors influencing the provision of memorable tourism experiences in social tourism programmes.
- Identify which tourism activities may be most suitable for the different segments of the social tourism market.
- Examine the vision and voice of children (e.g. children with strong financial constraints and children with disabilities) in terms of needs, desires, travel constraints and benefits obtained.
- Quantify the social and environmental impacts of social tourism programmes.
- Examine how social tourism may be implemented in developing countries.
- Ascertain whether it is possible to transfer social tourism systems adopted in developed countries to developing countries.
- Quantify economic benefits of social tourism initiatives to inform policy-makers of

the real benefits of these programmes in various domains (e.g. employment, household income, state revenues and health expenditures).

- Identify best practices in social tourism initiatives for all disadvantaged groups.
- Examine the real effects of social tourism trips on health and well-being, through interdisciplinary approaches, involving researchers with different knowledge; for example, in tourism, psychology and medical sciences.

References

Bélanger, C.É. and Jolin, L. (2011) The International organisation of social tourism (ISTO) working towards a right to holidays and tourism for all. *Current Issues in Tourism* 14(5), 475–482. DOI: 10.1080/13683500.2011.568056.

Diekmann, A., McCabe, S. and Ferreira, C.C. (2018) Social tourism: research advances, but stasis in policy. *Bridging the divide. Journal of Policy Research in Tourism, Leisure and Events* 10(3), 181–188.

Index

Page numbers in **bold** type refer to figures and tables.